MOVING TO MARS
DESIGN FOR THE RED PLANET

EDITED BY JUSTIN MCGUIRK, ANDREW NAHUM
AND ELEANOR WATSON

the DESIGN MUSEUM

Foreword   5
Justin McGuirk

Introduction   7
Andrew Nahum

IMAGINING MARS   13

IMAGINING MARS   17
Mike Ashley

ARRIVE   65

DESIGNING FOR THE MOST CHALLENGING   69
JOURNEY, EVER
Stephen Petranek

SARTORIAL THOUGHTS ON   93
TAILORING SPACESUITS
Anna Talvi

SURVIVE   125

MISSION CRITICAL: MARS MODERN   129
Fred Scharmen

CONTAMINATING THE RED PLANET   157
Lydia Kallipoliti

THRIVE   177

AN INTERVIEW WITH   181
KIM STANLEY ROBINSON
Justin McGuirk

THE WILDING OF MARS   205
Alexandra Daisy Ginsberg

Biographies   211
Index   212
Acknowledgements   219

Foreword
Justin McGuirk

Why would a design museum be interested in Mars? You could also phrase that question in reverse: why would those shooting for Mars be interested in design? The answer is the same: people. It is only recently that the space industry has been seriously considering – or, I should say, reviving – the idea of sending humans to the Red Planet. We have been successfully putting landers on Mars since 1976, but people are a different prospect. No space agency is going to send humans to Mars unless it can keep them alive and ensure tolerable living conditions for them, both on the voyage and on the surface of the planet. That factor, in the most literal sense, opens up a whole new world for design.

This book takes no unified position on the feasibility of that project. The reader will encounter the opinions of Mars boosters and Mars sceptics alike. But that does not change the fact that there is a considerable amount of design work being undertaken by space agencies, private companies, academics and practising designers of various stripes, all focused on the human experience of journeying to Mars and dwelling there for a period of time. This book, and the exhibition it accompanies, captures what it can of that work, as well as of new works commissioned specifically to make some of the challenges more tangible.

These challenges, which are considerable, fall largely into two phases. First is the voyage: keeping a crew safe and sane as they travel fifty-three million kilometres (thirty-three million miles) over seven months or more. In zero gravity, all the daily rituals – from eating and dressing to going to the toilet – need to be fundamentally rethought. The second phase is surviving on the surface, in supremely inhospitable conditions. It is not possible to have an unmediated experience of Mars. Astronauts will not be able to breathe the air or pick up a handful of regolith with their bare fingers. They would be enveloped at all times in artificial bubbles, protecting them against the temperature, the air and the radiation. Both their habitat and their suits are hermetically sealed worlds. In his essay, Fred Scharmen evokes that beautifully with the idea that leaving the habitat is just a small bubble peeling off a larger one. These bubbles are designed.

If we are serious about sending humans to Mars, then it is no longer just a scientific and technological challenge, but a design project. The human experience, and human survival, are paramount. Contained in these pages you will find architects, product designers and clothing designers considering how to address the many physical and psychological issues any putative crew would face. Their proposals are speculative at best. They are reaching towards what Kim Stanley Robinson, in this book, calls the material 'grammar' of a new world. Really this is a form of design fiction. But what better discipline to give nascent form to an emergent future?

Finally, some will ask why humankind would waste the mental energy and resources on getting to Mars when our home planet is in desperate need of saving. The editors of this book have enormous sympathy with that question. We are also sensitive to the fact that Mars itself is a contested site, and that human 'colonisation' is not necessarily a blessing. We offer no shoulds or shouldn'ts. But given that such a mission will most likely be undertaken, the question is: what can we learn from surviving on Mars that might help us survive on Earth? The extremes of the Red Planet demand new levels of lightness and efficiency, and better systems for extending scarce resources. Perhaps the very barrenness of Mars will help us avoid the same fate at home.

INTRODUCTION
Andrew Nahum

In 1996, the aerospace engineer and space activist Robert Zubrin published *The Case for Mars*. Subtitled 'the plan to settle the Red Planet and why we must', the book contained Zubrin's argument that the Mars venture was far more than an astronautical and scientific project. It was, he implied, a psychological and a spiritual necessity. Back in history, he suggested, the adventuring life on a new frontier – the old American West – refreshed and reinvigorated human culture. But, in his vision, 'the entire global civilisation based on ... humanism, science, freedom and progress will eventually die' unless democracy gets a shot in the arm from the example of a new frontier people. 'The Martians can show us the path away from oligarchy and stagnation.'[1]

NASA, mosaic of the Valles Marineris hemisphere of Mars, 2013.

Zubrin is one of the more zealous advocates of Mars exploration. The Mars Society he founded and the scheme being developed by Elon Musk's SpaceX together represent the most ambitious expressions of the Mars movement. However, many space scientists are more sceptical about the ease with which the environmental challenges of the planet can be met and the extent to which ambitious settlement schemes could 'live off the land'.

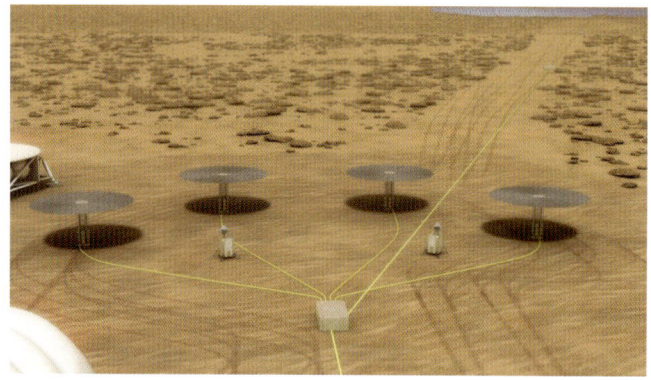

NASA, concept for a fission power system on the surface of Mars using four ten-kilowatt units, render, 2018.

So the mission being planned by NASA and the European Space Agency adopts a more conservative, incremental approach, with its *Orion* capsule resembling, in some ways, a bigger, more developed version of the Apollo craft that went to the Moon fifty years ago. It is intended initially to land just three astronauts on the surface for a stay devoted to survey, science and exploration.

For the 'strong' Mars movement, this is disappointingly unambitious, too slow, too old-fashioned and too expensive. But 'old-style' space engineering is, to a degree, quite conservative, because designers trust what works: fault prediction and verification for new spacecraft with many complex, interacting life-critical systems is hard. All machines fail sometimes.

Whatever the speed of Mars exploration, any project will be critically dependent on energy supply. According to one NASA study, a mission to land three astronauts for a sojourn of some months on the planet would need a power source of about forty kilowatts (kW). That would be enough for between fifteen and thirty homes in Europe or the United States, but on Mars that power is needed to keep just three people alive. Outside temperatures can hit minus sixty degrees Celsius at night, even on the equator, so part of the load is heating. It also goes to charge up their rover vehicles, their suits, power tools, communication systems and to extract pure-enough water from the ice frozen in the Martian subsoil.

Electricity is required to make breathable oxygen with a (still experimental) system that splits CO_2 extracted from the thin Martian atmosphere. The system would need to make liquid oxygen, too, for the rocket ride home.

Perhaps the key fact here is that this generating equipment weighs about ten tonnes – a demanding and expensive payload to launch from Earth. It could be in the form of solar panels, or it could be in the form of small nuclear-fission reactors, which appear to have an advantage in terms of mass.

This is one of the greatest challenges to spreading out over Mars: energy. It is probably feasible to launch enough ten-tonne power sets for a tiny population of explorers, but if thousands of settlers are to prosper there it would require thousands of tonnes of electrical power equipment, which is not going to be coming from Earth. Can these small generator sets – the 'founder generation' of Martian power plants – make more machines like themselves and create an expanding power network?

NASA, ISRU system concept for production and storage of oxygen and methane rocket fuel using carbon dioxide and water from the Martian surface, render, 2019.

Mars is rich in minerals. So can the first Martian settlers refine silicon to make new solar panels, locate and refine aluminium or copper ore to make the cables, or even mine uranium? How will the complex processes we use on Earth be reinvented on Mars? It's a process that space scientists call 'in situ resource utilisation' (ISRU) and it seems almost unthinkably hard, but perhaps that is to underestimate the power of human imagination and human ingenuity. After all, with Tesla, Elon Musk accelerated the introduction of the electric car by at least a decade, and his company, SpaceX, has developed booster rockets that have leapfrogged many launch systems from the world's well-established state-run national research agencies.

Why go at all, if it is so hard, and so dangerous? The astronomer Carl Sagan suggested that we should not attempt to settle Mars if life, any form of life, exists there – even tiny extremophile bacteria that may be deep underground. Others simply argue that colonising and exploiting another world 'when we have trashed this one' is intrinsically unethical. But can actions be unethical if there is no sentient life there to suffer harm?

In one countervailing and intriguing variation of this thought, Alexandra Daisy Ginsberg has explored an alternative future in which humans forbear to settle Mars but seed it with bacteria and plants. Her simulation of an evolutionary path that 'wilds' Mars by plant colonisation, followed by adaptation and evolution, is both an intriguing thought experiment and an artwork – a computer simulation running at over a million years per hour.

So where does the impetus for humans to explore Mars come from, at a time when many think that saving the environment on Earth should have far greater priority? Real futurists simply believe in old-style exploration because 'it's in our DNA'.

Or is the Mars project simply utilitarian? Not so long ago, Stephen Hawking argued that humankind should commit to Mars simply as insurance. Recently, another physicist, Michio Kaku, asserted that 'extinction is the norm. ... It is as inescapable as the laws of physics that humanity will someday confront some type of extinction-level event. ... we must leave the Earth or we will perish'.[2] His thought is echoed by Elon Musk, who has argued that if there's a third world war we want to make sure there's a seed of human civilisation somewhere else to bring it back and shorten the length of the dark ages.

HASSELL, 3D-printed habitat, render, 2019.
An inflatable habitat protected from the Martian
environment by a shield formed from dust and rock.

Maybe it is only a truly audacious research project like the Mars shot that can re-energise our technology, produce a more insightful ecological philosophy, foster a new sense of 'planetary management' and revolutionise our design thinking. One thing that is certain is that, for Mars missions and Mars habitats, everything will be designed in a new way, and with new constraints.

The hard-edged engineering solutions that took us to the Moon were endured by heroes who had spent their earlier lives strapped in the tiny cockpits of fighter aircraft. For the eight days or so it took to accomplish their missions, they adapted themselves to an uncompromising machine world.

But Mars voyagers will commit to live for years in new craft and new homes. These spaces demand a level of design that is much more focused on human needs and psychology. For example, the seven-month ride out there is, in itself, an arduous undertaking.[3] Earth-orbit astronauts have a timetable filled with scientific work and observations. Mars travellers are likely to have much more personal time. How should they fill it? Their psychological health will be crucial, and is rated by NASA as one of the major unknowns and major risk factors. With this in mind, the Design Museum has commissioned a 'dining table' for the journey from the designer Konstantin Grcic. Currently, space stations do have engineered structures that serve for eating meals and doing a variety of functional tasks. By contrast, Grcic has sought to create something that performs technically but is more pleasant, responding to the perception that the explorers will need a sociable space that encourages the valuable ritual of communal meals.

The surface of Mars, in the longer term, poses entirely new problems for housing, for clothing, for furniture and for recreation. Design factors include the lack of atmosphere, with less than one per cent of Earth pressure, extreme cold, high cosmic and solar radiation, and an extreme scarcity of Earth-produced construction materials. These constraints are producing an emerging Mars architecture using the Martian topsoil – the regolith – bound or stabilised in various ways to produce structures thick enough to give thermal and radiation protection, perhaps with an inflated membrane inside to retain air at a breathable pressure. And what will you wear? Outside, inevitably, some kind of pressure suit of the kind space exploration has made familiar, but planetary exploration really demands something more lightweight, more flexible and more human. For inside the habitat, designer Anna Talvi proposes garments that are both fitted and resistant – that stretch and restrain motion in a controlled way, so that every action demands a certain effort. Astronauts currently need to spend hours a day on exercise machines to try to counter muscle wasting in low gravity. Talvi's outfits are intended to make some of this protective exercise a constant and subliminal fact of life, like gravity on Earth.

Future Martian interiors provoke fascinating speculation. With Mars gravity at just over a third of that on Earth, there is tremendous scope for leanness and economy in design. Martian chairs, tables, beds – even shelves – will be entirely different to Earthly ones. How will future design on Mars evolve? Surely, after a few years of settlement, an extraordinary aesthetic will develop – a mix of supremely high-tech articles shipped from Earth at huge cost, furniture and utensils cannibalised from cast-off spaceship containers and parts or from cargo containers, and maybe other items carved from Martian rock or moulded from the regolith.

SEArch+ (Space Exploration Architecture), Mars X-House, crew quarters, 2019.

On Mars, energy is scarce, materials are precious, space is restricted. All this argues for a stripped-down life, conceived for the utmost economy in habitat, furniture, utensils and possessions. But it must still embody humane design and even 'spark joy'. Designing for life out there should teach us how to find real economy, but also how to extract real value – and real pleasure – from the resources we use on Earth.

1   Robert Zubrin, *The Case for Mars: the plan to settle the Red Planet and why we must* (New York: Simon & Schuster, 2011), p. 331 and elsewhere.

2   Michio Kaku, *The Future of Humanity: Terraforming Mars, Interstellar Travel, Immortality, and Our Destiny Beyond* (London: Penguin Books, 2018), 3.

3   Travel time to Mars varies between seven and twelve months according to the amount of fuel used and the orbital position of the planets at the time the journey is made.

IMAGINING
MARS

IMAGING MARS
Mike Ashley

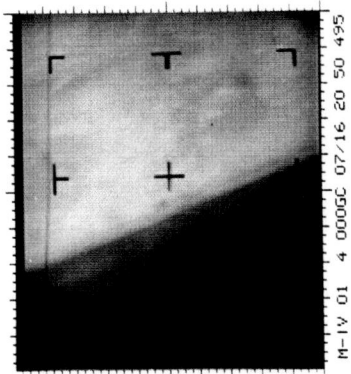

NASA, fly-by photograph of Mars'
surface taken by Mariner 4, 1965.

These days, thanks to the Martian rovers Curiosity and Opportunity, we can see photographs of the surface of Mars as if we were just metres away. But these pictures have only been achieved in the last twenty years, and it was not until 1965, with the fly-by of Mariner 4, that we had any close images of Mars.

Before then, even the best telescopes showed only a blurred surface, with markings that were open to interpretation and left a lot to the imagination. When the Italian astronomer and physicist Galileo turned his telescope on Mars in 1610, he saw no significant features, but at that time Mars was especially far from Earth. It was not until 1659, when Mars was closer (in 'opposition') that Christiaan Huygens produced a vague map of the planet. He tried again in 1672, when he depicted the southern polar ice caps, possibly already observed by Giovanni Cassini. But Mars was too distant to understand with the weak telescopes of the time. When the French philosopher Bernard de Fontenelle compiled his groundbreaking treatise on the plurality of worlds in 1686, he was dismissive of Mars, deeming it unremarkable.

A century later, with the observations by William Herschel in 1784, opinion grew that Mars may well be remarkable after all, and possibly the only planet in the solar system likely to harbour life. Herschel believed he had detected seasons on Mars and that the dark areas might be seas. In 1860, Emmanuel Liais suggested that these dark areas were not seas but vegetation.

The ancient view of Mars had been influenced by its red, blood-like colour and its odd movement through the heavens. Mars was considered strange, even sinister, and some speculated that it might be where our spirits go after we die — a heaven or a hell. Now views were changing, and the idea that there might be intelligent life on Mars came into focus with the observations of Giovanni Schiaparelli in 1877. His map showed large dark areas connected by lines that he called *canali*, meaning 'channels'. This was translated as 'canals', and that one word was enough for reporters to speculate that, as canals are artificial waterways, there must be intelligent life, with canals of that scale implying an advanced civilisation.

Schiaparelli did not himself advocate genuine canals, but it soon became a subject of much debate among scientists. The one who did most to promote the idea of life was Percival Lowell, a rich American who had built his own observatory in Arizona. Lowell set down his thoughts in *Mars*, published in 1895, and two later books which developed in detail his theories about Martian life. He speculated that because of the lesser gravity Martians would be three times taller than humans, resulting in strong muscles and greater efficiency of movement. He also believed any such life would be highly intelligent.

Though many astronomers argued against them, Lowell's speculations caught the public's imagination. This was further stimulated by HG Wells' novel *The War of the Worlds*, serialised in *Pearson's*

Magazine in 1897, with dramatic illustrations by Warwick Goble. Wells proposed a dying Mars, whose intelligent, warlike inhabitants sought to take over the Earth. Although Wells' Martians had weak, amoeboid bodies, they had built formidable walking war machines armed with death rays. The Earth was at their mercy.

Wells' novel was immensely influential. It went through many editions and sequels were written by others, with illustrations by many artists. Although he did little to describe Mars itself, Wells' image of an advanced Martian race able to conquer Earth remains vivid more than a century later. Orson Welles' 1938 radio adaptation was so powerful it allegedly caused panic in New Jersey, the location of the fictional Martian landing. Two film adaptations and a recent TV version continue to keep Wells' vision alive, along with such films as *Invaders from Mars* (1953) and the spoof *Mars Attacks!* (1996).

The books by Wells and Lowell encouraged the public to believe that the Martians might be observing Earth. There was even a march for piano, *A Signal from Mars* composed in 1901 by Raymond Taylor, with sheet music that showed Martian telescopes trained on Earth. Popular magazines ran features on Mars, often heavily illustrated, including artwork by Henri Lanos in the French magazine *Je sais tout* (*I know all*) in 1906, showing fairy-like winged Martians finding recreation in observing daily life on Earth. Wells speculated further on the nature of Martian life, describing beings that were covered in feathers or fur, nine or ten feet (2.7-3m) tall, with several feet or a prehensile tail, and more intellectually advanced than humans.

Warwick Goble, illustration from the original serialisation of *War of the Worlds* by H G Wells published in *Pearson's Magazine*, 1897.

The idea of a warlike Martian race persisted but became merged with the vision of Mars as a dying world, striving to survive, which gave it a more romantic image. American writer Edgar Rice Burroughs, the creator of Tarzan, chose Mars for a series of adventure novels that started in the pulp magazines in 1912 with 'Under the Moons of Mars' (published in book form as *A Princess of Mars*, 1917). His hero, John Carter, is transported to Mars where the lesser gravity gives him superior strength and agility. Burroughs had little interest in the scientific reality of the planet. His delight was in using an alien environment for heroic adventures and romance – Carter falls in love with the Martian princess Dejah Thoris. Burroughs' novels gave rise to the genre of planetary romance, later exploited by Leigh Brackett, Marion Zimmer Bradley, Lin Carter and others, portraying a fantasy form of Mars with minimal scientific basis.

The horrors of the First World War, however, meant that the public demanded hope for the future, and Mars came to be viewed as an alternative, more idealised Earth. It was portrayed beautifully in the remarkable Danish film *Himmelskibet* (1918). When released in the United States as *A Trip to Mars* in 1920, the film came with a poster depicting the spaceship *Excelsior* that proclaimed 'You can now make reservations'. As one of the first post-war movies, *Himmelskibet* preached a message of love and harmony between Earth and Mars, presenting the latter as a Garden of Eden.

Orson Welles broadcasts his radio show *The War of the Worlds* at CBS Radio, 1938.

The red colour of Mars provoked other thoughts, especially in Russia. In the novel *Red Star* (1908), Alexander Bogdanov wrote of the planet as a socialist utopia. Alexei Tolstoi's *Aelita* (1923) portrays the angst of a capitalistic Mars ruled by a Martian elite, again featuring a princess. In 1924, *Aelita* was made into a film with bizarre expressionistic Martian sets. Mars as a dying world provided a potent setting for Lao She's *City of Cats* (1932, also known

Henri Lanos, illustration for
an article entitled 'The Call
of Another World' in *Je sais
tout* (*I know all*), 1906.

Himmelskibet (*A Trip to Mars*),
directed by Holger-Madsen, film
poster, 1918.

as *Cat Country*), a scathing satire of Chinese society in which a declining, corrupt and complacent civilisation, dependent on opium, is open to invasion.

The American inventor and publisher Hugo Gernsback, who launched the first science-fiction magazine in 1926, accepted the possibility of Martian canals and speculated about how they could have been built – using super-scientific machines – in his serial 'The Scientific Adventures of Baron Münchausen' in 1915. He described how the environment on Mars influenced the evolution of Martians, with a huge chest and lungs to capture oxygen from the rarefied air, large ears to hear in the tenuous atmosphere, and retractable eyes and nose to protect against the cold.

The Second World War brought a harsh reality to the romantic view of Mars, captured evocatively in Ray Bradbury's stories assembled as *The Martian Chronicles* (1950). These followed the gradual colonisation of Mars, and the extinction of native life on the planet. While paying homage to the romance of Old Mars, Bradbury recognised, in the post-nuclear world, humanity's need for outward expansion to establish itself on other worlds in case Earth became inhospitable. Mars might yet prove a lifeboat.

The imagination, having passed from anxiety to romanticism to idealism, now turned to realism – especially after 1947, when it was discovered that the atmosphere was almost entirely carbon dioxide. Wernher von Braun, who had invented the V-2 rocket during the war, described how Mars could be colonised in his book *The Mars Project* (1953); it was chosen as one of the fifty best books of the year by the American Institute of Graphic Arts. He considered in detail the problems and possibility of interplanetary flight. His work was picked by *Collier's Weekly* in 1954, with special features on 'Can We Get to Mars?' and 'Is There Life on Mars?', beautifully illustrated by Chesley Bonestell. Bonestell's work inspired many further designs, including Patricia Cullen's striking covers for Patrick Moore's adventure series, which began with *Mission to Mars* (1955), and John Richards' covers for the British science-fiction magazine *Authentic*, which depict the first landing on Mars via its moon Phobos.

Probes were sent to orbit Mars by the Soviet Union and the United States with mixed success since 1962, but it was not until 1076 that the first Viking probe landed on the planet and began a sustained programme of photography and experimentation. It soon became clear that Mars had no intelligent life forms, a hope that David Bowie had questioned in his song 'Life on Mars?' (1971).

Writers turned their attention to adapting Mars for humans, a process called 'terraforming'. Carl Sagan considered the ethics of this in *The Cosmic Connection* (1973), stating that, if it were discovered that Mars had any indigenous organisms that would be endangered by planetary re-engineering, then it should not be terraformed. He nevertheless believed that by releasing the water locked in the polar ice caps it might be possible to create more Earth-like conditions within a few hundred years. Following Sagan's lead, Kim Stanley Robinson produced the definitive work on terraforming Mars in his fiction trilogy *Red Mars* (1992), *Green Mars* (1993) and *Blue Mars* (1996).

By the start of the twenty-first century, the accumulated scientific evidence showed a Mars that looked, to all intents and purposes, like a barren wasteland, dotted with craters, deep valleys, high mountains and no canals, but with evidence that there may once have been rivers and

*Authentic Science Fiction
Monthly*, cover, 1954.

lakes. The rocks are mostly basalt covered with an iron oxide dust, which gives the planet its red colour, and which winds can whip up into a sandstorm, obliterating the surface. Yet the possibility that it might at one time have supported life remains a tenacious hope. In *Mars* (1992) and *Return to Mars* (1999), Ben Bova used the latest scientific data to demonstrate both the difficulties confronting any attempt to colonise Mars and the thrill of exploring an ancient world.

The chances of surviving on Mars were realistically portrayed in Andy Weir's *The Martian* (2011), in which an astronaut, left for dead on the Red Planet, uses all his scientific training to survive until he is rescued. The novel is both a remarkable demonstration of the determination of the human spirit and a powerful message of hope for the future of humanity in venturing beyond Earth. Four centuries after Galileo, Mars still captures our imagination.

Clay cuneiform tablets showing planetary observations for Mars and Mercury during the early fourth century BC, Late Babylonian.

Calyx krater, c. 430 BC. Urn featuring a depiction of the five planets visible from Earth with the naked eye as boys diving into the sea.

Dresden Codex, fourteenth century. Thirty-nine-leaf manuscript depicting ritual and divination calendars as well as calculations of planetary movements.

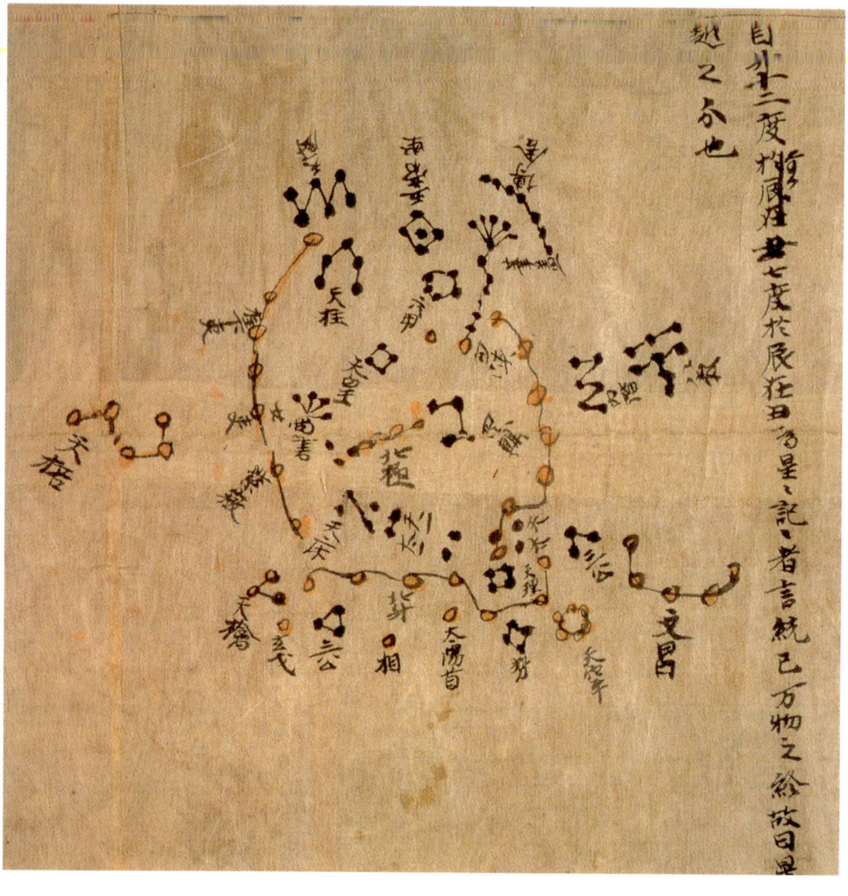

Dunhuang Star Chart, c. 700 AD. The earliest known map of the night sky.

Mangala, Hindu god of Mars, painting, c. 1725-50. The planetary deity Mars (or *Mangala*)
flanked by Kārttikeya and Lakshmi.

Mars is a constant presence in the sky, glowing red and clearly visible with the naked eye. Its changing movements have fascinated humans for millennia, forming an integral part of Ancient Greek, Sumerian, Hindi, Mayan and Persian astronomy. The planet's fiery colour has meant it is often associated with blood and war, as with the Greek god of war, Ares.

Andreas Cellarius, *Scenographia
systematis Copernicani*
(*Scenography of the Copernican
world system*), 1708.
Astronomical chart devised
by Nicolaus Copernicus in
the early sixteenth century.

In 1543, Nicolaus Copernicus transformed
our understanding of the universe with
his revolutionary heliocentric model,
which placed the Sun, rather than the Earth,
at the centre of our solar system. His
theory was partially verified through his
observations of Mars, whose retrograde
motions had long confounded astronomers.
Johannes Kepler, a follower of Copernicus,
later used his observations of Mars to
confirm that the planets move in elliptical,
rather than circular, orbits around the Sun.

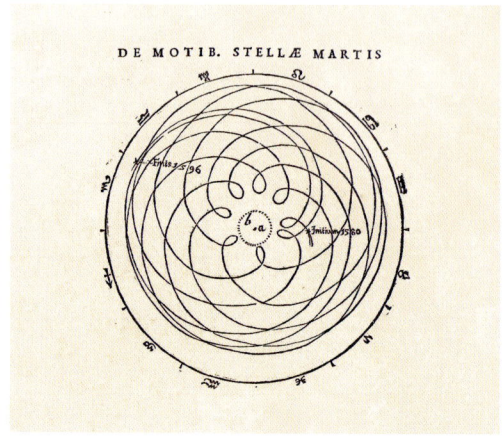

Johannes Kepler, diagram of the geocentric
trajectory of Mars from his book *Astronomia
Nova (New Astronomy)*, 1609.

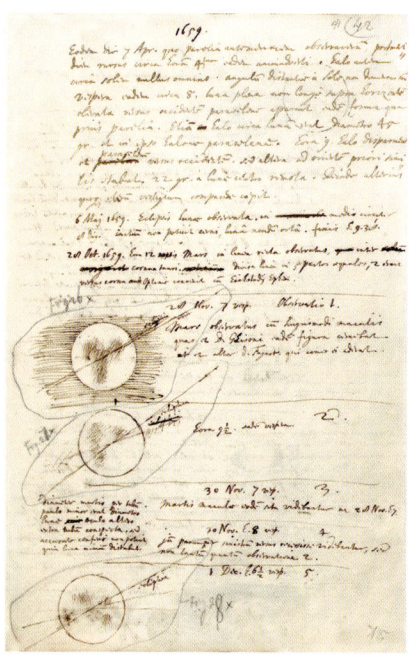

Christiaan Huygens, Mars observations
including a sketch of the surface of Mars
depicting Syrtis Major, 1659.

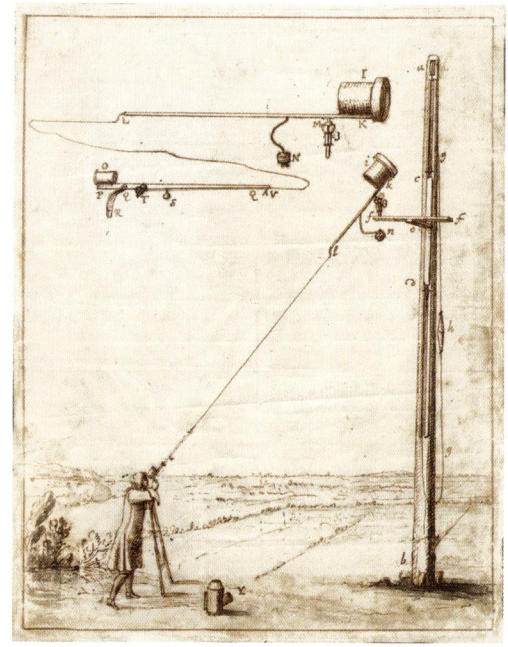

Christiaan Huygens, drawing of an aerial
telescope, c. 1684.

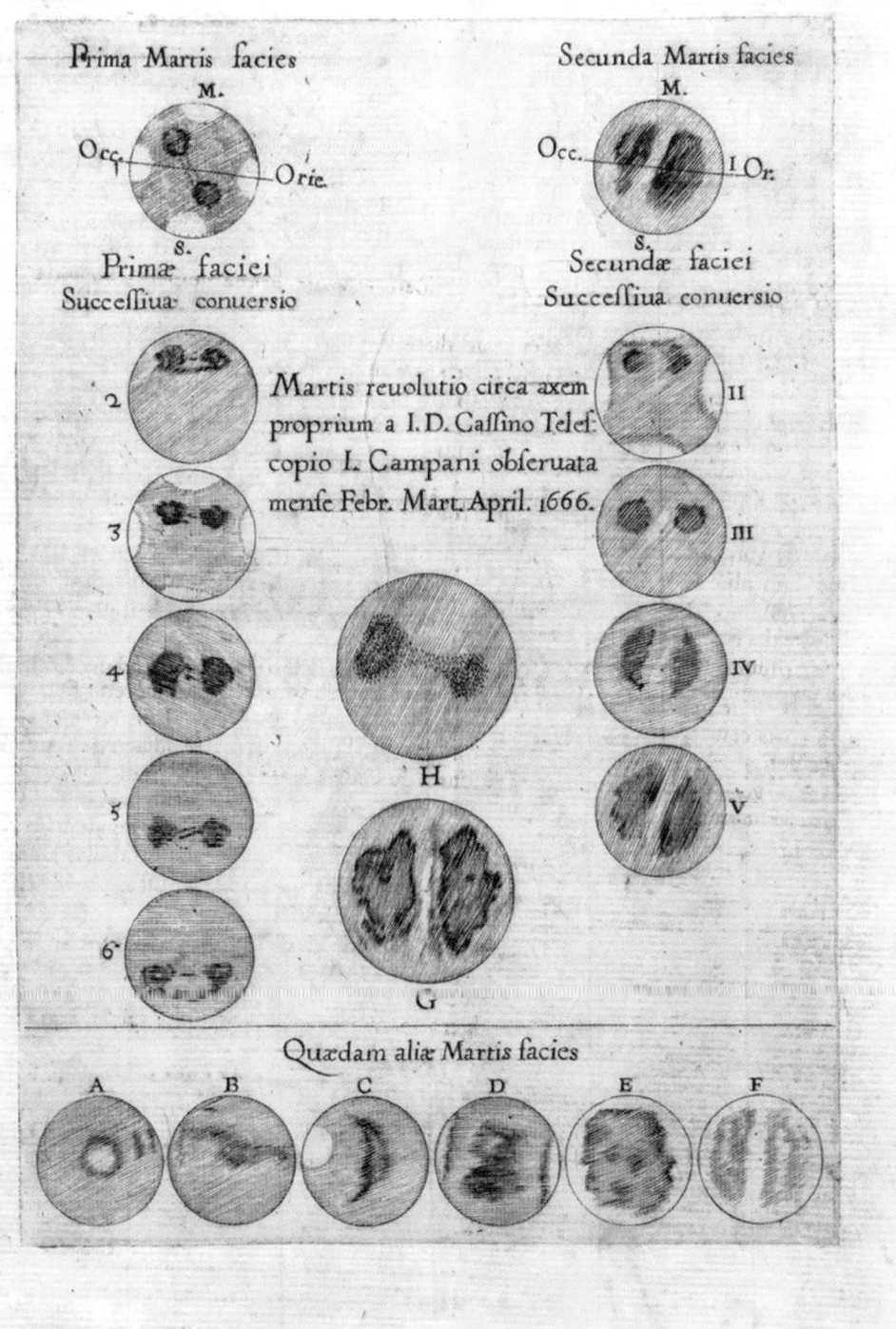

Giovanni Domenico Cassini, drawings of Mars from his book *Observations in Bologna of the Rotation of Mars Around its Axis*, 1666.

William Herschel, reflecting telescope, c. 1775. A 2m-tall (7-ft) Newtonian telescope.

William Jones, orrery, c. 1798-1830. A mechanical model of the solar system used to predict the positions and alignments of planets.

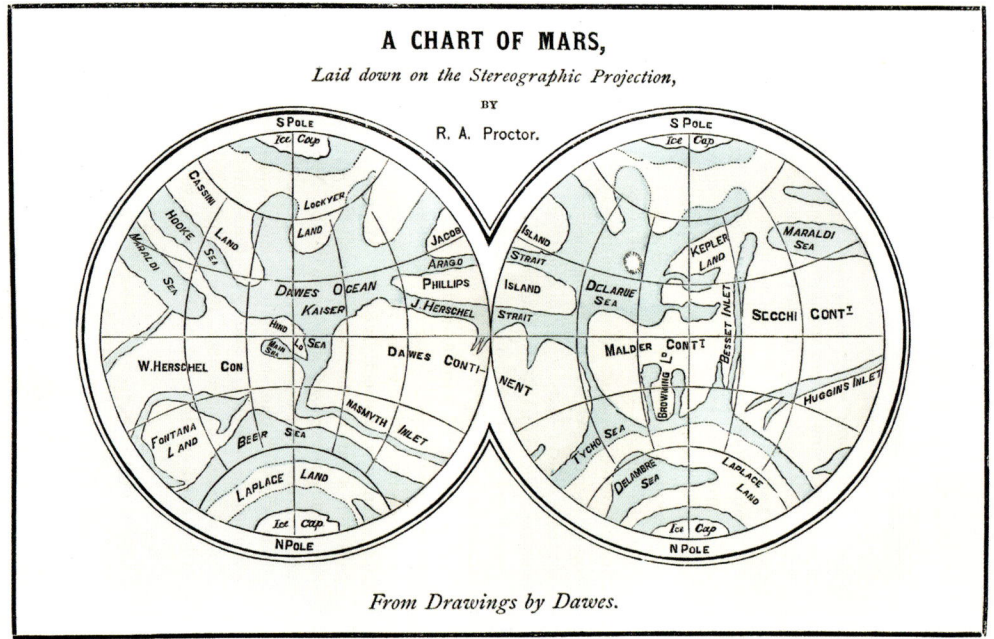

Richard A Proctor, map of Mars from his book *Other worlds than ours: the plurality of worlds studied under the light of recent scientific researches*, 1871.

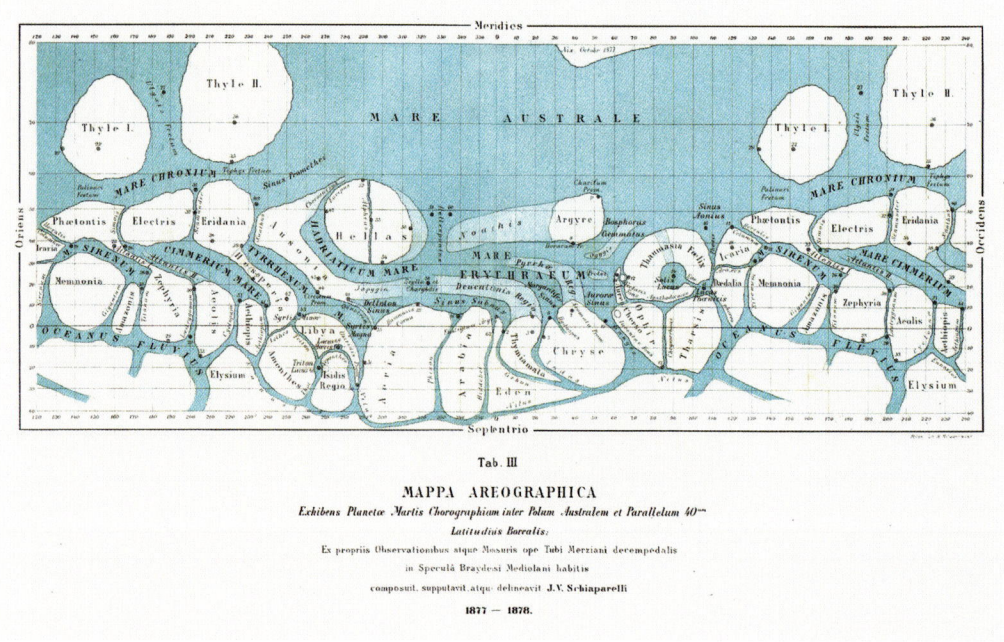

Giovanni Schiaparelli, map of Mars from his book *Osservazioni di Marte* (*Observations of Mars*), 1878.

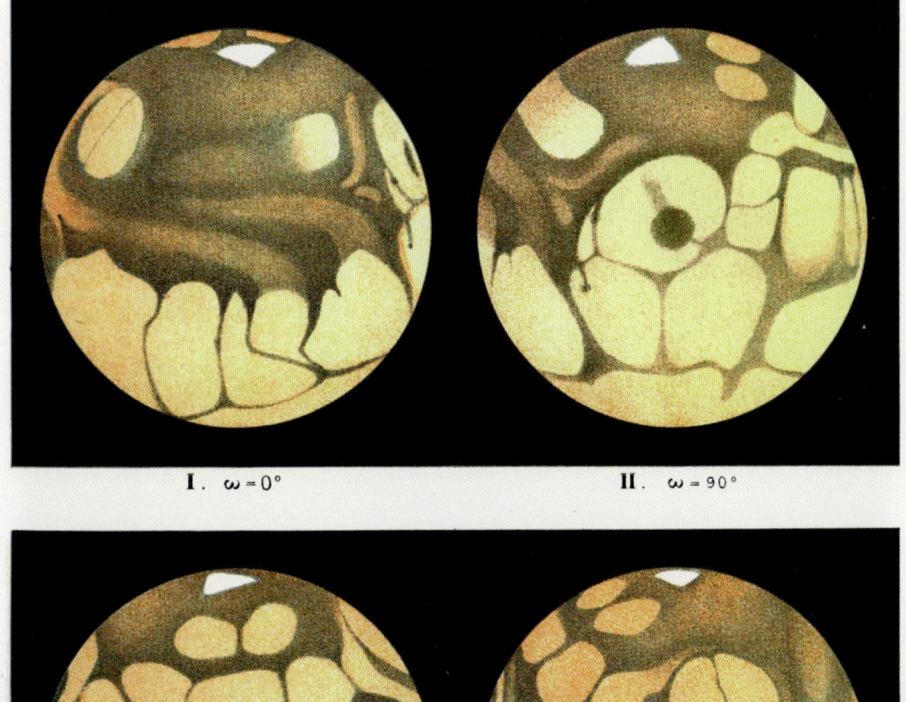

Giovanni Schiaparelli, map of Mars from his book *Osservazioni di Marte* (*Observations of Mars*), 1878. The surface of the planet as seen through a telescope when Mars and Earth were in 'opposition', or as close to one another as their orbits allow.

Achille Beltrame, drawing of Giovanni Schiaparelli featured on the front page of *La Domenica del Corriere* (*The Sunday Paper*), 1900.

Tab. IIII.

Hemisphærium Martis Australe

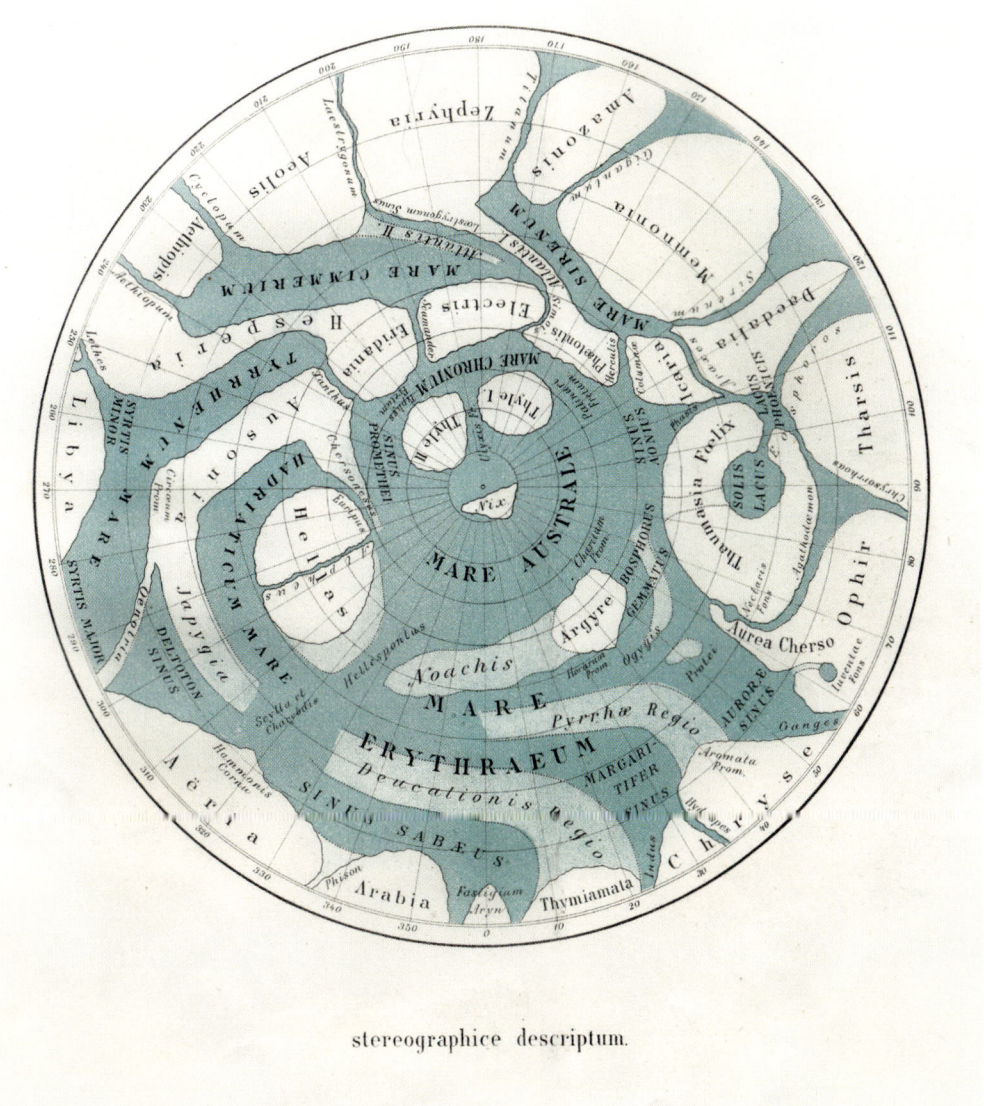

stereographice descriptum.

Giovanni Schiaparelli, map of a Mars pole from his book *Osservazioni di Marte* (*Observations of Mars*), 1878.

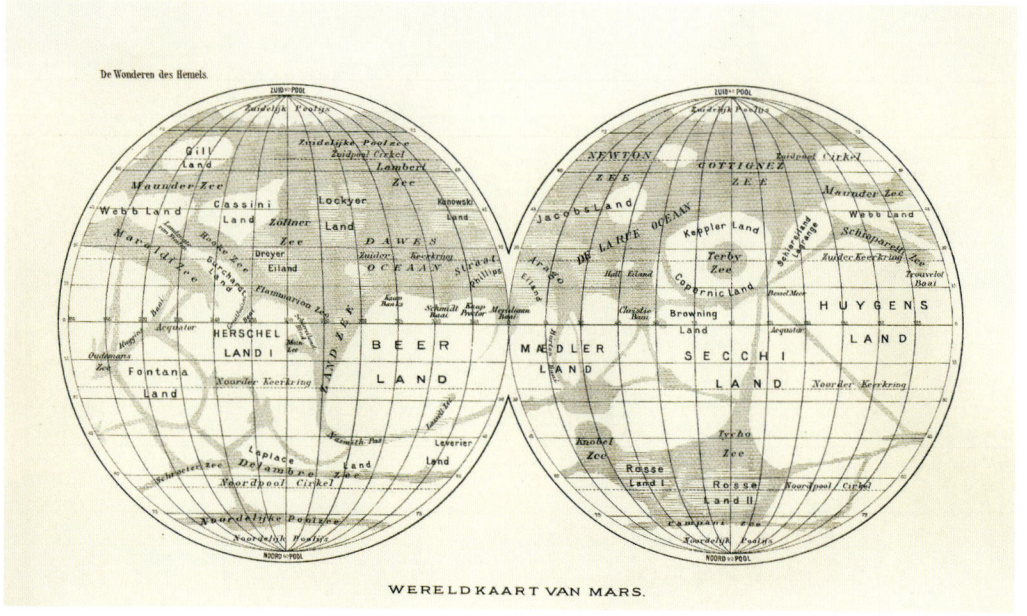

Camille Flammarion, map of Mars from the German edition of his book *Les Merveilles Céleste*
(*The Wonders of the Heavens*), 1870.

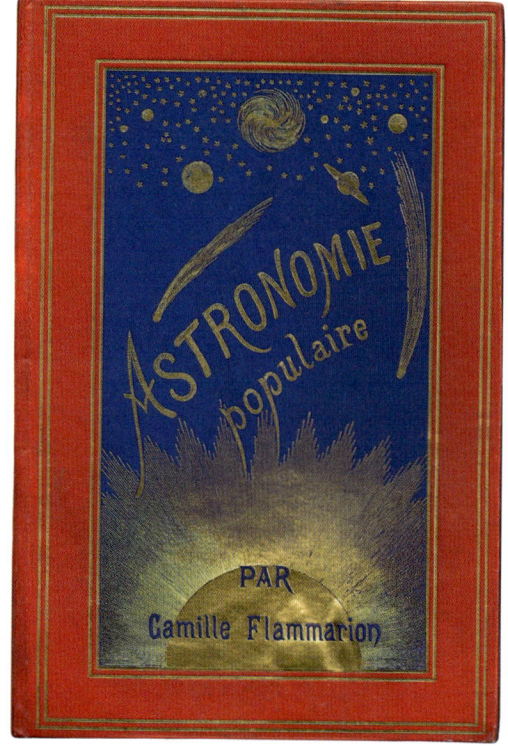

Camille Flammarion, *Astronomie Populaire*
(*Popular Astronomy*), book cover, 1880.

CAMILLE FLAMMARION A SON OBSERVATOIRE DE JUVISY   n̊ H.27.6

Épreuve obtenue avec l'objectif du "CONGRÈS"

Camille Flammarion at the eyepiece of his 240mm (9.5") Bardou refractor telescope in his
Juvisy observatory, mid-1880s.

Percival Lowell at the 610mm (24") Clark Telescope, c. 1914.

Born to a prominent Boston family, Percival Lowell dedicated his considerable wealth to the observation of Mars. He was inspired by Camille Flammarion's popular science book *La Planète Mars* (*The Planet Mars*), as well as the numerous maps of Mars produced by Italian astronomer Giovanni Schiaparelli.

Adamant that the dark markings on the planet were canals created by intelligent beings, Lowell published a series of books about Mars, fuelling the 'canal craze' that has defined our perception of the planet to this day.

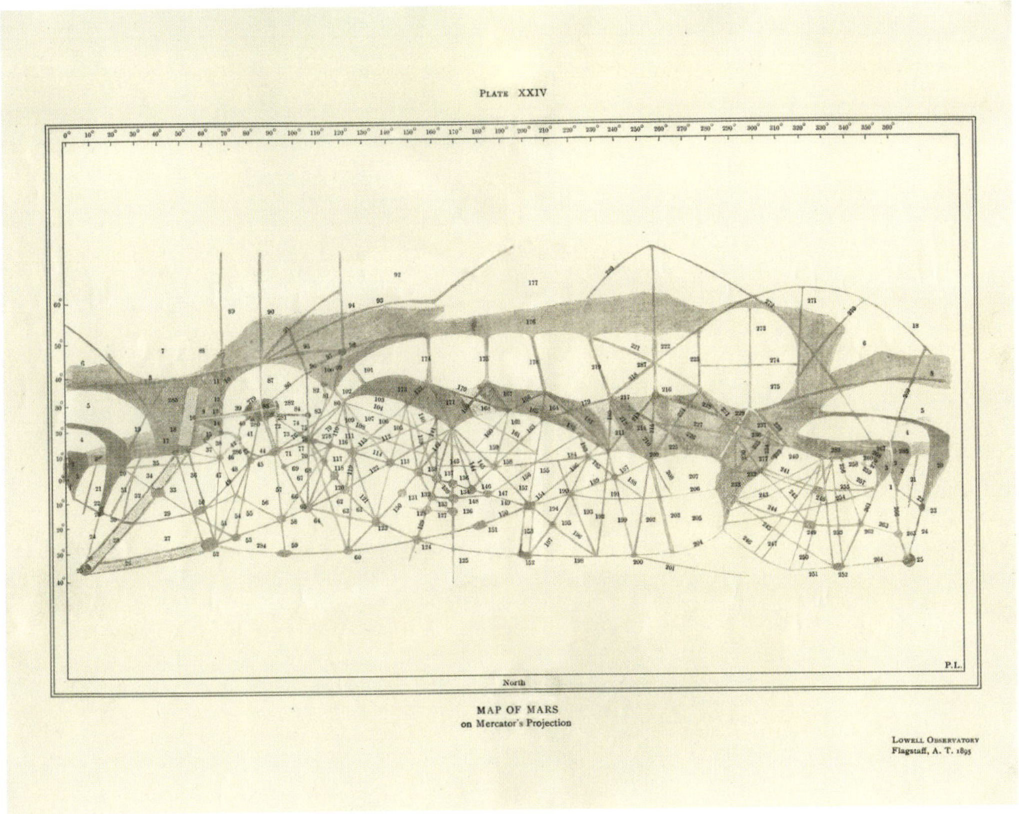

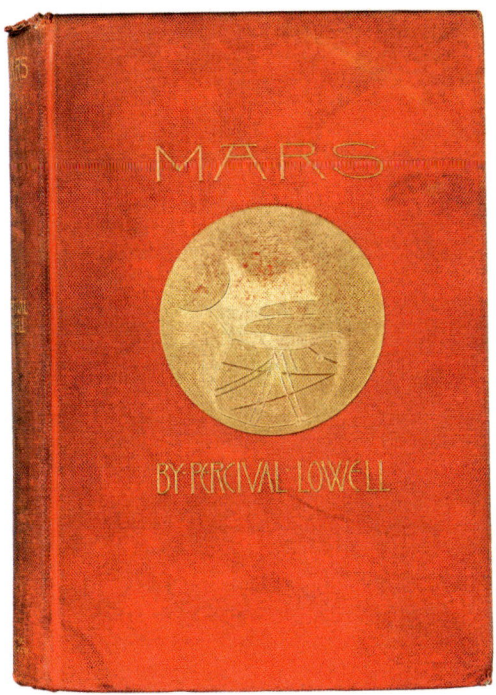

Percival Lowell, map of Martian canals, from his book *Mars*, 1895.

Percival Lowell, *Mars*, book cover, 1895.

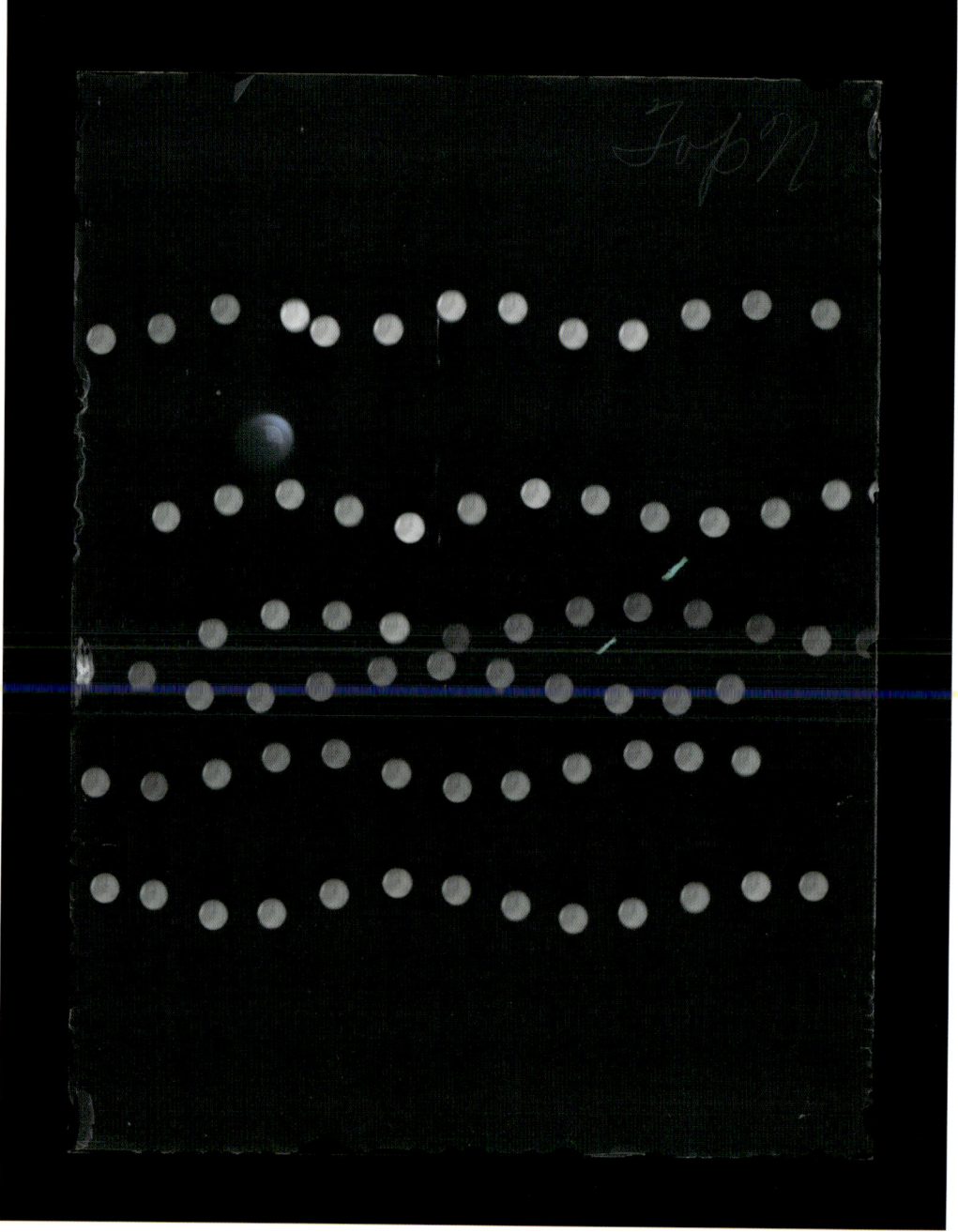

EC Slipher, photographic glass plate showing the movements of Mars, 1907.

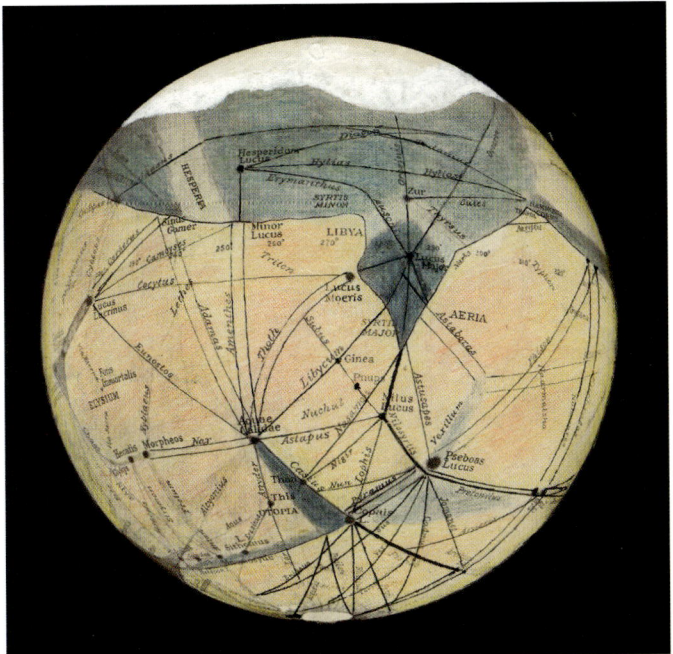

Percival Lowell, colour drawing of Mars, 1905.

The 610mm (24") Clark Telescope Dome in Tacubaya, Mexico, 1896. The Clark Telescope was shipped to Mexico from its usual home at the Lowell Observatory in Arizona to allow Lowell to view Mars in opposition.

Henrique Alvim Corrêa, poster announcing the publication of the Belgian
edition of HG Wells' *The War of the Worlds*, 1906.

Frank R Paul, illustrated cover of *Amazing Stories* magazine, 1927.

The War of the Worlds is the most influential example of Martian science fiction. First serialised in 1897 in *Pearson's Magazine*, it was one of the first stories to detail a conflict between humankind and an extraterrestrial race. The story gained widespread notoriety when Hollywood director Orson Welles staged a live reading of the story on CBS Radio in 1938, which listeners mistook for a news report.

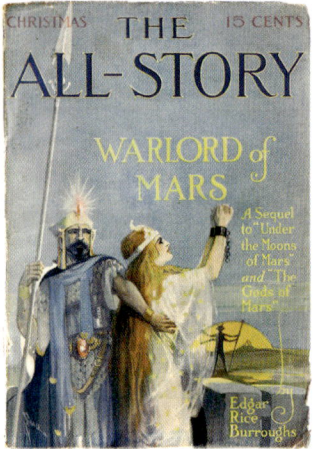

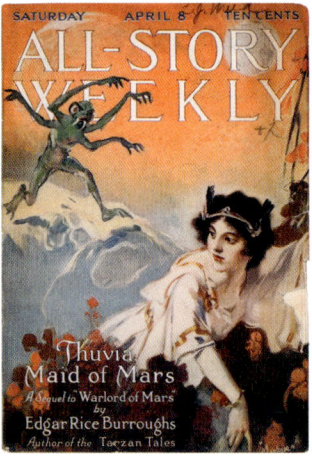

Fred W Small, cover of *All-Story* magazine featuring an illustration of Edgar Rice Burroughs' serialised story 'Warlord of Mars', 1913.

PJ Monahan, cover of *All-Story Weekly* featuring an illustration of Edgar Rice Burroughs' serialised story 'Thuvia, Maid of Mars', 1916.

J Allen St John, cover of Argosy *All-Story Weekly* featuring an illustration of Edgar Rice Burroughs' serialised story 'The Chessmen of Mars', 1922.

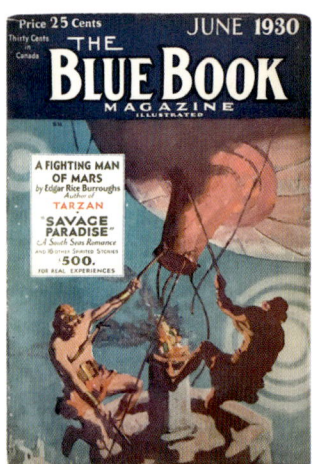

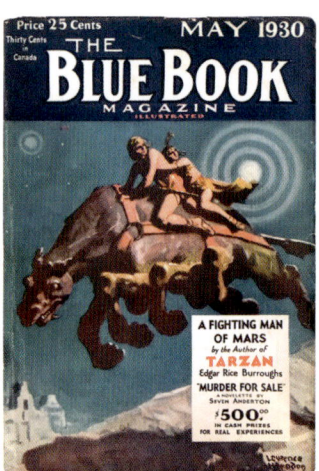

Laurence Herndon, cover of *Blue Book* magazine featuring an illustration of the serialised story 'A Fighting Man of Mars', 1930.

Aelita: Queen of Mars, directed by Yakov Protazanov, film poster, 1924.

Aelita: Queen of Mars, directed by Yakov Protazanov, film still, 1924.

Invaders from Mars, directed by William
Cameron Menzies, film posters, 1953.

Carl Urbano, *Destination Earth*, film stills,
1956. Created by John Sutherland, this short
promotional cartoon features a Martian
sent to Earth to uncover the secret behind
America's prosperity – namely, petroleum
and the free market.

Davis (John Richards), cover of *Authentic Science Fiction Monthly* magazine, 1953.

Davis (John Richards), cover of *Authentic Science Fiction Monthly* magazine, 1954.

Frank R Paul, illustrated cover of *Wonder Stories* magazine, 1936.

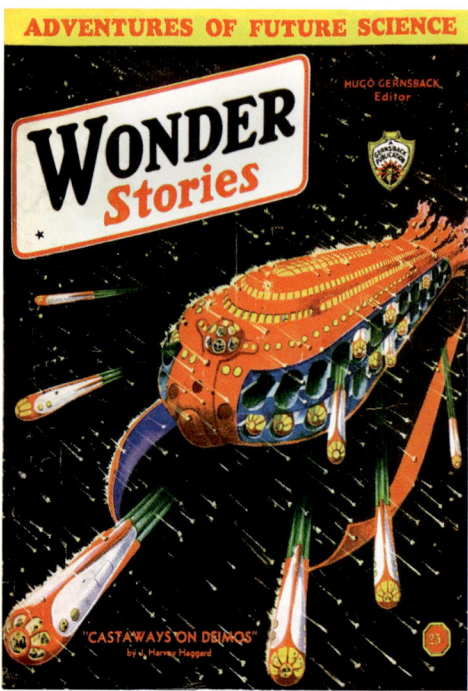

Frank R Paul, illustrated cover of *Wonder Stories* magazine, 1933.

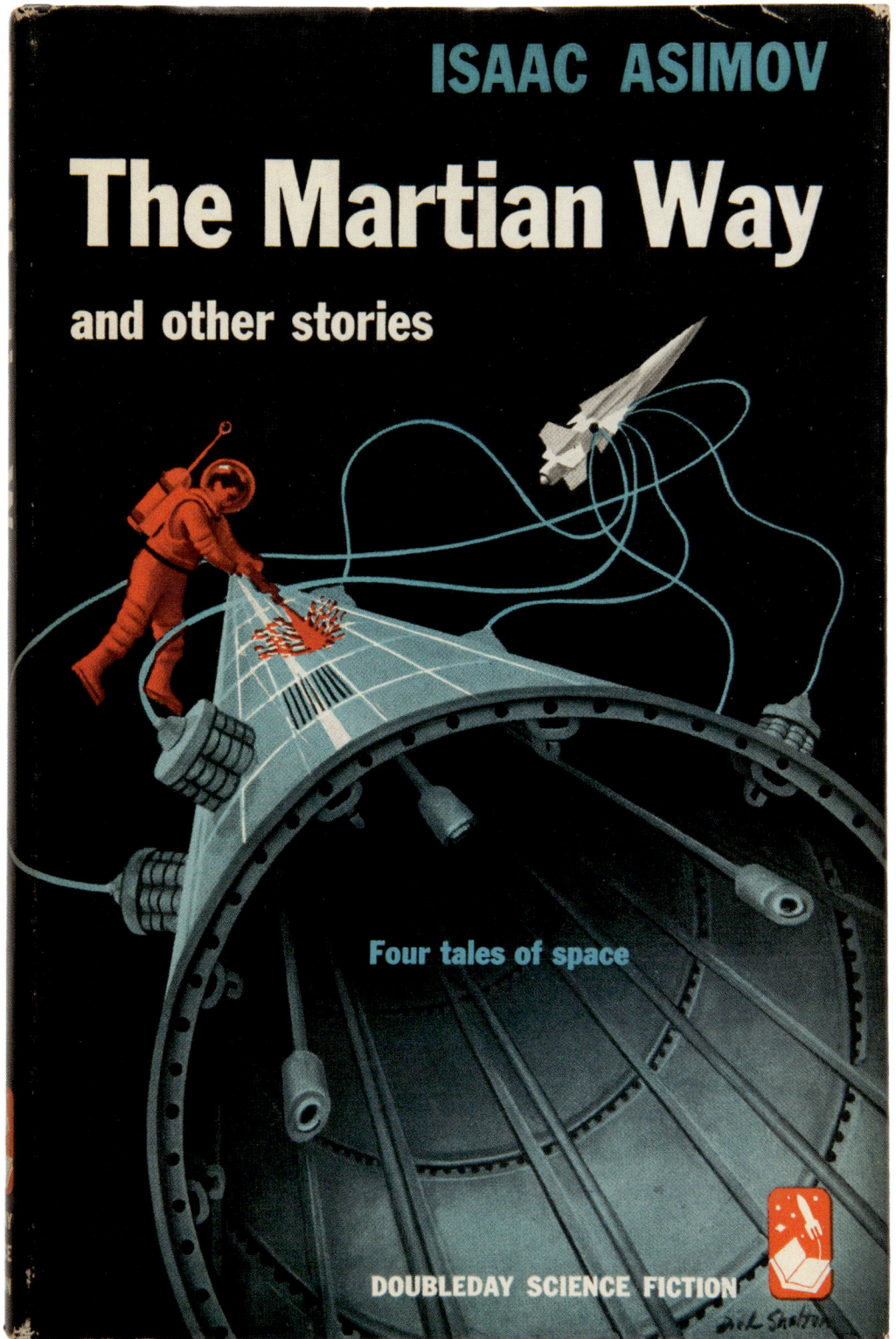

Richard Shelton, cover of Isaac Asimov's *The Martian Way and other stories*, 1955.

Tekhnika Molodezhi (Technology for the
Youth), magazine illustration, date unknown.

Tekhnika Molodezhi (Technology for the
Youth), magazine illustration, date unknown.

Tekhnika Molodezhi (Technology for the
Youth), magazine illustration, date unknown.

Tekhnika Molodezhi (Technology for the
Youth), magazine illustration, date unknown.

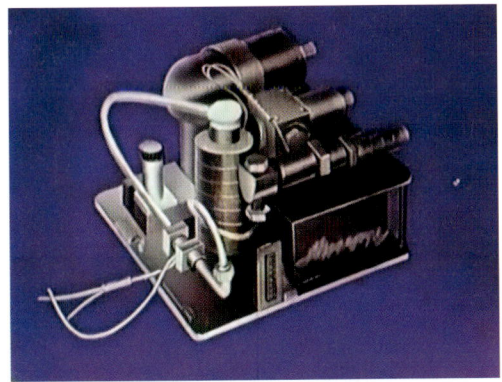

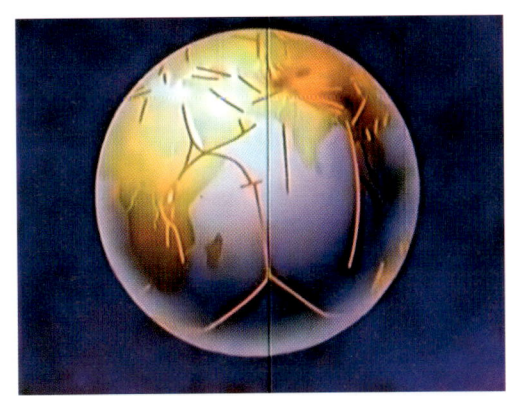

Mars, directed by Pavel Klushantsev,
film stills, 1968.

Nebo Zovyot (*The Sky is Calling*), directed
by Valery Fokin, film poster illustration
by Y Tsarev, 1959.

Despite the near-total failure of their
Mars missions, the Red Planet was a
recurring feature of Soviet science fiction.
The earliest – and best-known – example
is the feature-length film *Aelita: Queen
of Mars* (see p. 43). Produced in 1924
and based on Alexei Tolstoy's novel of
the same name, the film sees a young Soviet
engineer travel to Mars where he takes
part in a workers' revolution against the
planet's oppressive leaders.

Invaders from Mars, directed by Tobe Hooper, film poster, 1986.

Total Recall, directed by Paul Verhoeven, film still, 1990.

Total Recall, directed by Paul Verhoeven, film poster, 1990.

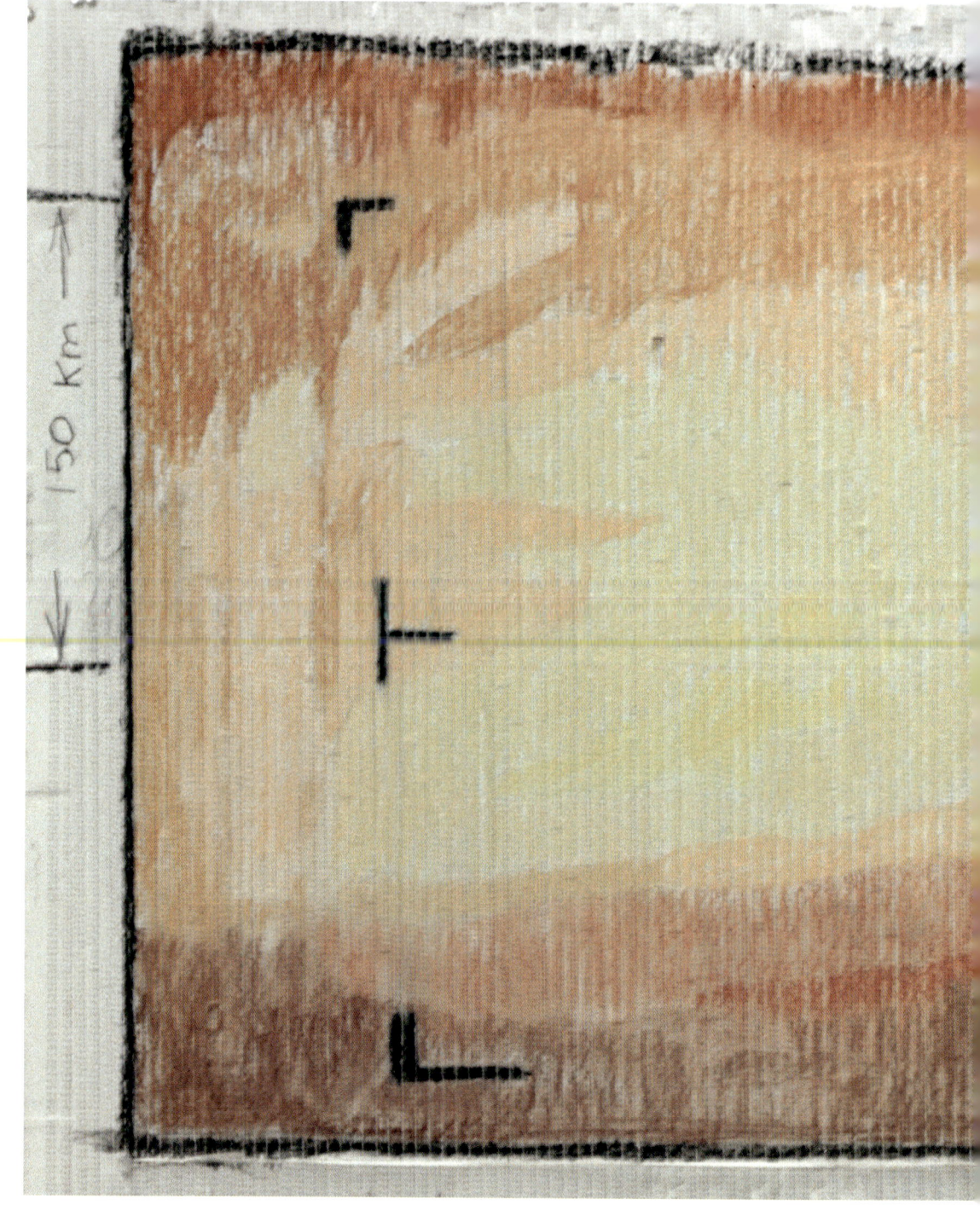

NASA, scientist hand-colouring the first digital image of the surface of Mars,
sent by Mariner 4, 1965.

Mariner 4 was a watershed moment in the history of Mars observations. It managed the first successful fly-by of the planet and captured twenty-one clear photographs of the Martian surface, cratered and seemingly devoid of life. Each photograph took six hours to be sent back to Earth, and the scientists had to wait a considerable period of time for the image data to be computer processed. In their impatience, they used a set of pastels to colour in a numerical print out of the raw pixels, meaning that the first digital image of Mars was actually hand-drawn.

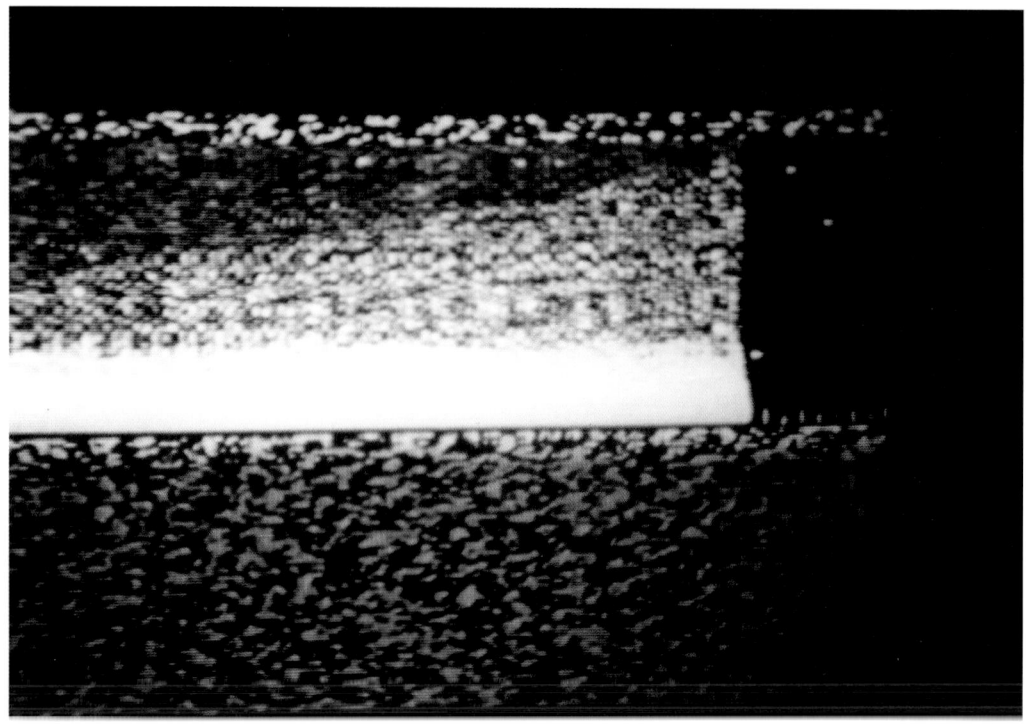

Soviet Mars Program, the first image of the surface of Mars, sent by the Mars-3 lander, 1971.

Soviet Mars Program, Mars-3, 1971. The Mars-3 lander was the first spacecraft to touch down successfully on Mars, but failed only a few seconds after landing.

NASA, the first clear photograph taken from the surface of Mars, sent by the Viking 1 Lander
shortly after it touched down, 1976.

Patricia 'Patsy' Conklin assembles a mosaic of Mars' surface from photos taken by
unmanned space probe Mariner 9, 1972.

NASA, the Sojourner rover on the surface
of Mars photographed by cameras on-board
the Mars Pathfinder, the spacecraft that
transported it to the planet, 1997.

NASA, technicians work on the Mars
Exploration Rover 2 (MER-2) ahead of its
launch at NASA's Kennedy Space Center, 2003.

NASA, the first true colour photograph of Mars, taken by the Mars exploration
rover Spirit, 2004.

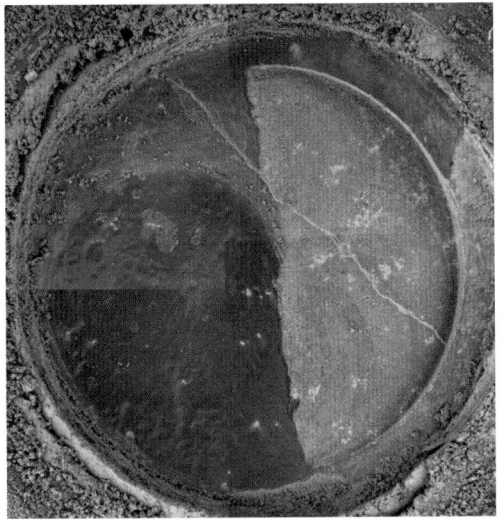

NASA, a mosaic of photographs providing
a microscopic view of the drilled surface
of Martian rock named 'Mazatzal', taken
by the Spirit rover, 2004.

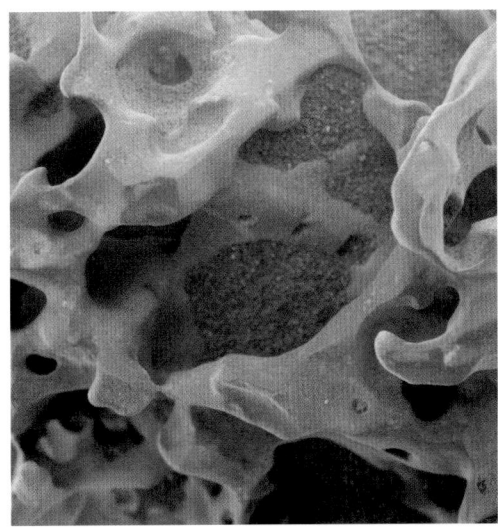

NASA, microscopic image showing the erosive
effect of winds on hardened lava, taken by
the Spirit rover, 2006.

NASA, compilation images showing the movements of Mars' moons Phobos and Deimos as captured
by the Spirit rover, 2005.

NASA, a billion-pixel resolution image taken by the Curiosity rover at the Martian site known as 'Rocknest', 2012.

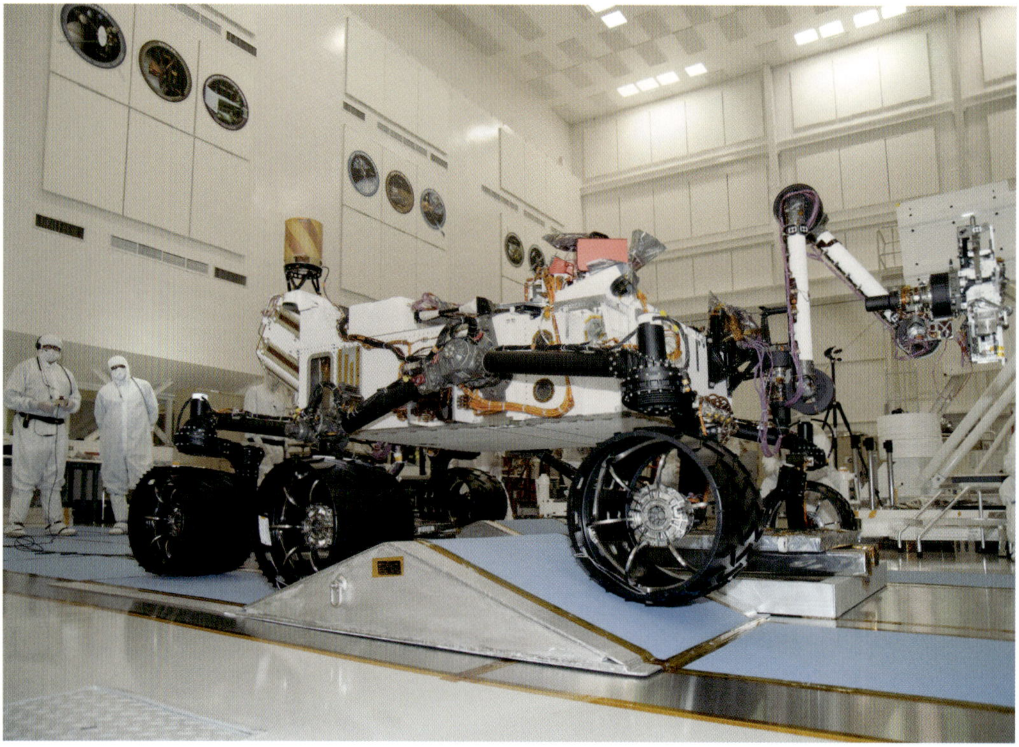

NASA, Curiosity rover on a test drive, 2010.

NASA, a composite self-portrait formed of numerous images taken by the Curiosity rover
of itself using its Mars Hand Lens Imager, 2015.

NASA, Curiosity's view of a Martian dune after crossing it, 2014.

Indian Space Research Organisation (ISRO), Mars Orbiter Mission probe, also known as Mangalyaan, 2013.

ISRO, true colour image of Mars taken by the Mars Orbiter Mission probe, 2014.

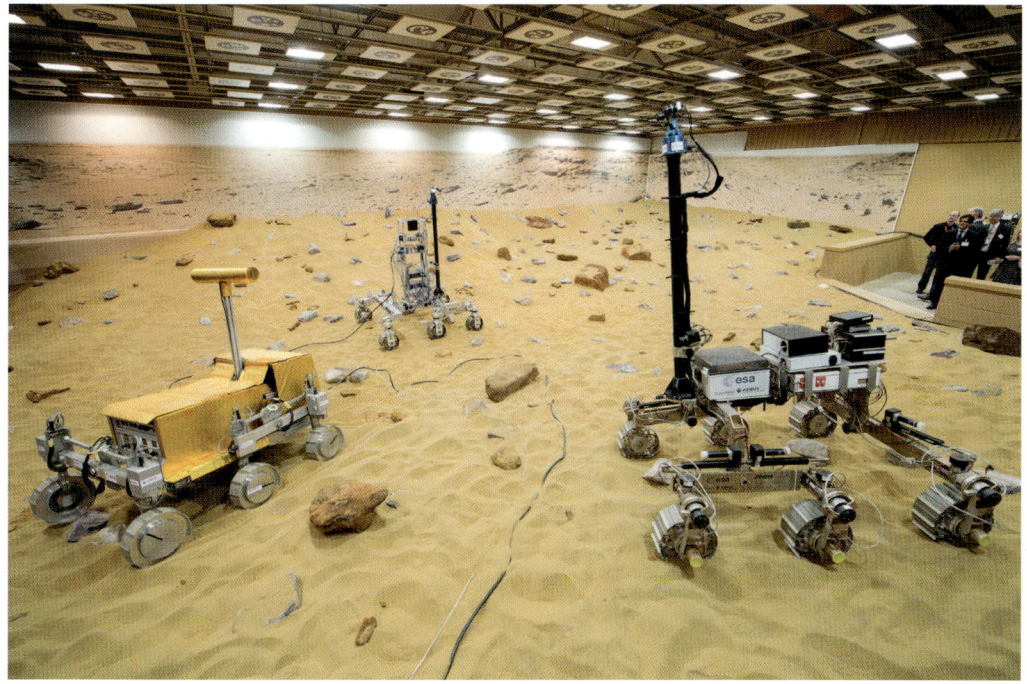

ESA, ExoMars Rovers 'Bridget', 'Bryan' and 'Bruno' on the Mars Yard test bed area at Airbus
Defence and Space in Stevenage, UK, 2014.

Since the first successful fly-by in 1964, geological conditions. The planet is
fourteen orbiters and four rovers have currently being charted by NASA's Curiosity
successfully been sent to Mars, mapping rover, which has sent back over half
the entire planet's surface and gathering a million images since landing in 2012.
data about Mars' atmosphere, climate and

NASA, an artist's impression of the InSight lander on Mars, render, 2018. Short for 'Interior Exploration using Seismic Investigations, Geodesy and Heat Transport', InSight is designed to give the Red Planet its first thorough 'check up' since it formed 4.5 billion years ago. It will look for tectonic activity and meteorite impacts, study how much heat is still flowing through the planet, and track Mars' wobble as it orbits the Sun.

ARRIVE

DESIGNING FOR THE MOST CHALLENGING JOURNEY, EVER
Stephen Petranek

The Moon is a mere 400,000 kilometres (250,000 miles) from Earth, about ten times the distance around our equator, and although the achievement of reaching its surface in 1969 stands as arguably the most inspiring event of the twentieth century, a manned landing on Mars would be exponentially more magnificent. That's because Mars is about 1,000 times farther away from Earth, an extraordinary 400 million kilometres (250 million miles). And it would take eight months to get there, as opposed to the three-day trip to the Moon. Simply keeping people sane during the voyage to Mars may be the greatest design challenge ever faced.

The father of modern rocketry foresaw the problem three-quarters of a century ago. In 1948, Wernher von Braun, the brain behind the Nazi V-2 rocket who was whisked out of Germany after the Second World War to work for the Americans, wrote a ninety-one page manual called *Das Marsprojekt* that was later published in English as *The Mars Project* (1953). It was a technical exploration with intricate calculations for exactly how humans could get to the Red Planet. All of it is still valid today.

To von Braun, the maths was easy; the hard part was how to make people comfortable and happy as they ventured farther and farther from home. Instead of a few brave souls on a dangerous voyage in a single spaceship, he insisted on ten rockets travelling as a fleet, carrying seventy people. He even designed a shuttle to ferry passengers from rocket to rocket, allowing social engagement and readjustment of the human mix. To keep people from constant space sickness, he figured out how to induce artificial gravity by tethering pairs of spaceships together with a cable and spinning them around each other like two yo-yos connected by a string.

Von Braun had dreamed of voyaging to Mars since he was thirteen, and his creation of the massive Saturn V rocket that got humans to the Moon was just a stepping stone. When the Apollo programme was winding down in the early 1970s, he campaigned vigorously for NASA to follow up by landing humans on Mars, which he said could be done by 1985.

To reduce costs, von Braun scaled down his original ideas. He said a Mars-shot could use a modified version of the Saturn V, carrying perhaps five astronauts in a larger Apollo-like craft, and that only three rockets might attempt the trip together. That vision is not much different from how NASA is planning to land humans on Mars in the 2040s, considerably more than fifty years after the end of the Apollo programme. In fact, the entire NASA Mars mission is a throwback. The *Orion* capsule that sits atop NASA's new Space Launch System (SLS) rocket, itself remarkably similar to a Saturn V, has been called 'Apollo on steroids'.[1] Several years ago, when I asked SpaceX CEO Elon Musk if he could see any technological advances between the *Orion* capsule and the Apollo capsule, he was quick to answer: No.[2]

NASA, Apollo 13 and Saturn V, 1969.

Meanwhile, competition to get to Mars has heated up, thanks to the singular corporate mission of SpaceX to build a sustainable city there. To Musk, sustainable means one million people, a number necessary to build an infrastructure that frees Martians from Earth support systems.

To accomplish its mission, SpaceX is building the largest rocket system ever to be launched from Earth – the Starship and its Super Heavy booster. This is not a vehicle designed for a few astronauts to collect rocks from another planet and then return home. It is the airline of the future. A Starship will be as reusable as a Boeing 747, and it will carry 80 to 100 people in comfort and ease for months through an environment so harsh it challenges the edges of technology and psychology.

Each SpaceX Starship will be 9 metres (30 feet) in diameter and 55 metres (180 feet) long. The design borrows from von Braun, including his insight that any spaceship headed to Mars has to leave from Earth's orbit, not its surface, because that would require too much rocket and too much fuel. So the thirty-nine-storey system has two sections: the reusable booster and the second-stage Starship. After the booster rockets out of Earth's deep gravity, the stages separate and the Starship, which is part spaceship and part rocket, fires into orbit. There it docks with a fuel tanker previously launched by the same reusable booster. After refuelling, it blasts off for Mars.

SpaceX, Starship, render, 2018.

The Starship will contain forty cabins, each with a window, lighting, electricity and connection to an intranet. Walls will be movable, so passengers can expand their private space to include others. SpaceX plans to land two cargo versions of the Starship on Mars in 2022, each carrying a payload of 100 tonnes of supplies for a later human landing. Musk hopes that four additional Starships will head for Mars in 2024, two with cargo and two with Earthlings aboard.

To get a million people to Mars, SpaceX is prepared to build 1,000 completely reusable Starships over the next thirty years. Therefore, by 2050, a fleet of 80,000 people could leave for Mars every two years, when the planets are aligned in such a way as to make the shortest trip. If there is enough demand to build this fleet, the population of Mars could increase by 400,000 just from 2050 to 2060.

Although these numbers are barely digestible, the concept of getting any group of people to spend eight months confined in a small stainless-steel can without killing each other may seem more far-fetched. Musk has acknowledged the problem by insisting that the voyage to Mars be 'fun' for passengers.[3] The idea is to create the opposite of the Apollo moonshot capsule, which was akin to the world's smallest submarine, with tiny windows and a single chair per passenger – the only perch for sitting, sleeping, defecating and eating. SpaceX designers know that one of the first requests of early astronauts on space stations was a table where they could share meals. Starship will offer communal spaces for both recreation and meals.

Depending on your reference point, the Starship is spacious, with 800 square metres (8,880 square feet) of pressurised cabin, at least as voluminous as the interior of an Airbus 380, which typically carries 550 passengers, and larger than the entire International Space Station (ISS). With two or three to a cabin, private space will be precious, and out-of-the-box design thinking will be critical. Designer Raymond Loewy, who consulted on the plans for *Skylab*, America's first space station, and made more than 3,000 design recommendations to NASA from 1967 to 1973, said that something as simple as colour – the more of it the better – can be important to mental health. Although Loewy designed everything from astronaut clothing to space scooters, he is remembered most for insisting that *Skylab* have windows – a luxurious six of them. Astronauts came to love staring out of those windows.

SpaceX has paid attention. From the first sketches, its designers filled the entire nose cone of the Starship with a honeycomb of windows, through which passengers can access the best view of the universe humankind has ever been offered – a constant gallery of visual wonderment. They also will serve the intense bonding that is likely to occur as passengers watch Earth recede to Carl Sagan's pale-blue dot and a reddish beacon ahead grow slowly larger. The visual experience will create unending awe, somewhat like living inside a telescope. As any ocean sailor can attest about a clear night at sea, the number of stars visible will crowd out the blackness of space with light.

There will be no sense of motion or movement for most of the journey, and stars will not seem to move through the sky. The experience will be one of eerie suspension, a silent and peaceful glide through nothingness. 'The great thing about going to space is there is no friction, so once you are out of the atmosphere, it is smooth as silk – no turbulence, nothing. There is no weather, there is no atmosphere,' Musk has said.[4]

SpaceX, Starship cutaway render, 2018.

In contrast, the short trip from Earth to orbit will be truly frightening: raucous cacophony from the Raptor engines, high G-loads that push towards unconsciousness, and vibrations that make the spacecraft seem as if it is shaking apart. Hours later, another short sequence of strapped-in terror will come after refuelling, when the engines ignite to blast off for the long coast to Mars. They will fire for no more than a few minutes. Then there will be weeks on end to contemplate the meaning of a one-way trip to the unknown.

The realities of day-to-day living on the Starship will require some getting used to. Astronauts on the ISS often comment about never getting tired of weightlessness, but sleeping in an actual bed is a dream that will not be fulfilled until voyagers arrive in Mars' meagre gravity. Instead, passengers will zip themselves into a sleeping-bag-like enclosure so they can't bump into things as they move during sleep. Chairs are almost useless in space, yet indispensable for landing and taking off. Although passengers will enjoy sitting around a table to share meals, they must strap themselves into chairs with seat belts. Other than sleeping, the most time-consuming part of each day must be exercising. Designers will need to employ every trick they can think of to make daily exercise a form of entertainment because, even when astronauts on a space station exercise several hours a day, they lose about one per cent of their bone mass per month.

Like passengers on a cruise ship, Starshippers will be required to practise emergency drills. About four times a day, the Sun spits out huge masses of high-energy particles that could easily kill or sicken everyone aboard. Because the Sun is a sphere and these coronal mass ejections can shoot out in thousands of different directions, the likelihood of crossing paths with them is small, but the result is too significant to ignore. Each ship will have a solar-storm shelter, which everyone must crowd into. For extra shielding, water can be pumped through the walls, ceiling and floor of the storm shelter, providing extra protection.

Astronaut Dan Burbank works out on the Combined Operational Load Bearing External Resistance Treadmill (COLBERT) on the ISS, 2011.

Water itself cannot be made in space and is very expensive to transport because it weighs a kilogram per litre (eight pounds per gallon). So passengers on the Starship will only borrow the water they consume each day. All fluids will be completely recycled for reuse, as they are on the ISS. Even water vapour expelled by breathing will be captured and recycled. Filtering systems must eliminate CO_2 expelled by breathing too. No one will take aboard more than two changes of clothing – the weight costs are too high. Clothes will be made of fabrics that shed dirt, perspiration and stains. High-speed super-efficient spin machines will be used to wash clothes once a week. Toilet tales from space travellers abound, and it is safe to say that nobody who has used the clunky vacuum-assisted devices designed for weightless environments has liked it.

Musk, who is fond of saying that people living on Mars will need 'pizza joints' and that 'Mars should really have great bars',[5] will have to decide if a Starship bar is a good idea. Limited alcohol consumption might be a welcome social lubricant, though it is supposedly forbidden on the ISS.

The end of the journey will bring one more terror, as the spacecraft shakes violently while aerobraking into the Martian atmosphere, slipping and leaning at an awkward angle towards the surface, with its windows side up to expose its entire 30-metre (180-foot) bottom as a heat shield. An unsettling red-hot glow will surround the craft. Then comes a slamming jolt, as the engines fire again, bringing the ship ever more slowly to the surface. A last burst of engine thrust leaves the Starship tail down, landing directly on the surface, resting on pads built into its three small delta wings aft. After one small step onto the Mars surface, humans will leap into an interplanetary future.

1   Mike Griffin, quoted in 'NASA News Conference with Mike Griffin: Exploration Systems Architecture Study (Transcript)', SpaceRef (20 September 2005), http://www.spaceref.com/news/viewsr.html?pid=18122 [Accessed 3 June 2019].

2   Stephen Petranek, *How We'll Live on Mars* (New York: Simon & Schuster, 2015), 25.

3   Quoted in 'Elon Musk envisions "fun" but dangerous trips to Mars (Update 4)', Phys.org (27 September 2016), https://phys.org/news/2016-09-spacex-ceo-musk-mind-mars.html [Accessed 3 June 2019].

4   Quoted in Catherine Clifford, 'Here's what it will be like to travel to Mars in Elon Musk's spaceship', CNBC.com (29 November 2017), https://www.cnbc.com/2017/11/29/what-it-will-be-like-to-travel-to-mars-in-elon-musks-spaceship.html [Accessed 3 June 2019].

5   Quoted in Natalie Jarvey, 'SXSW: Elon Musk Teases Trips to Mars by Early 2019, Reveals thoughts on HBO's "Silicon Valley"', The Hollywood Reporter (11 March 2018), https://www.hollywoodreporter.com/news/sxsw-elon-musk-trips-mars-by-early-2019-thoughts-hbos-silicon-valley-1093649 [Accessed 3 June 2019].

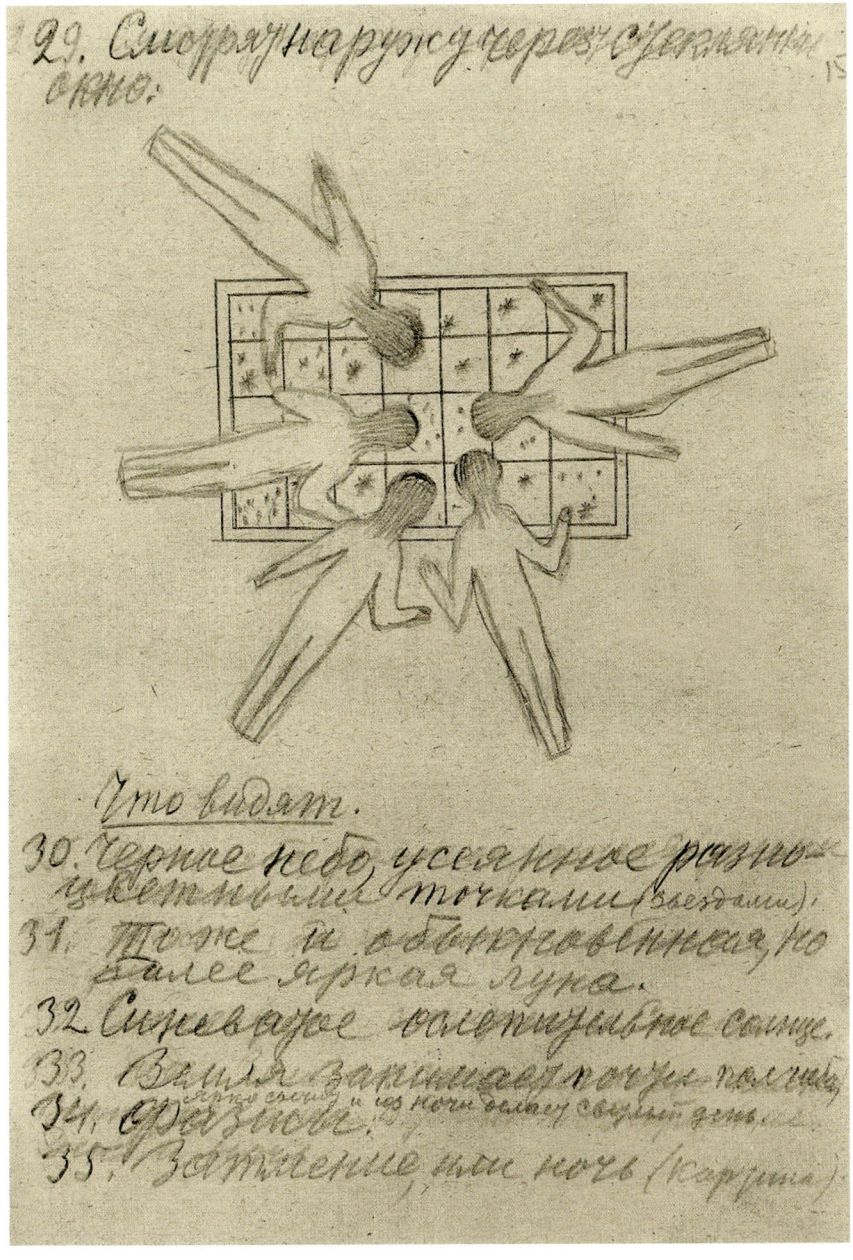

Konstantin Tsiolkovsky, sketch depicting astronauts floating at zero gravity from his *Album of Cosmic Journeys*, 1933.

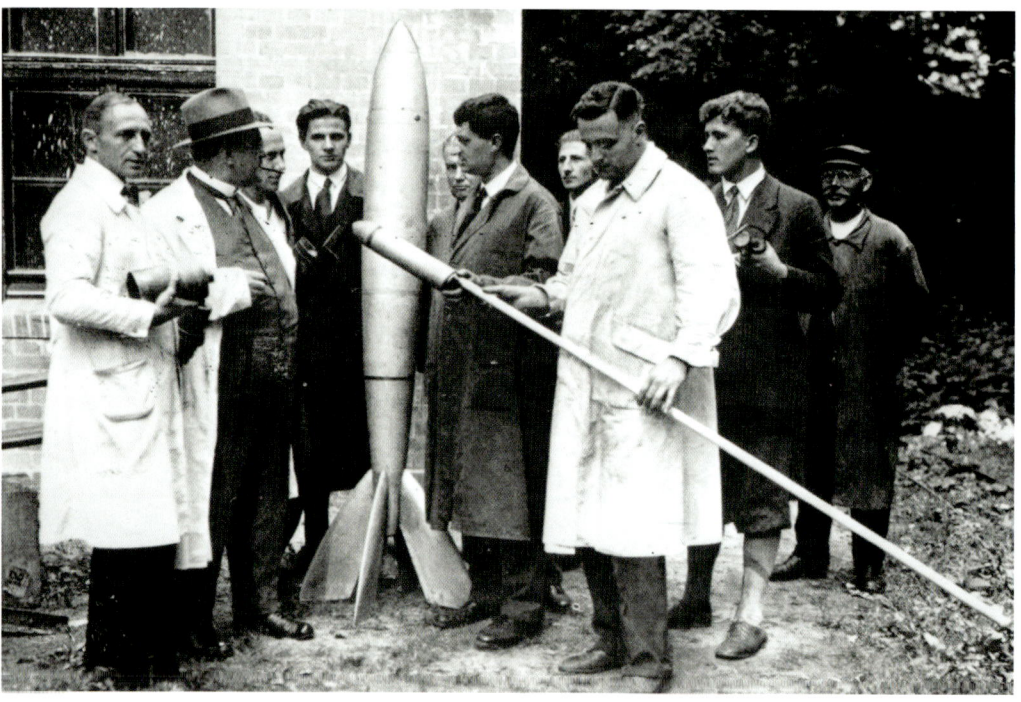

A young Wernher von Braun (second from the right) with other members of the German
Spaceflight Society, 1930.

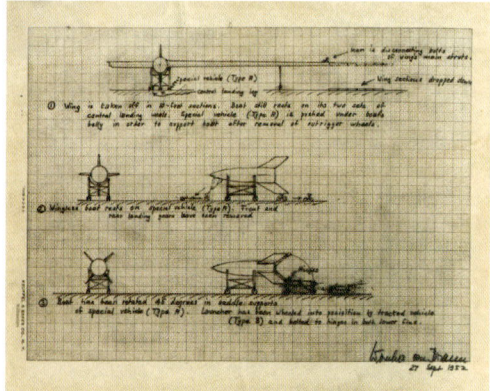

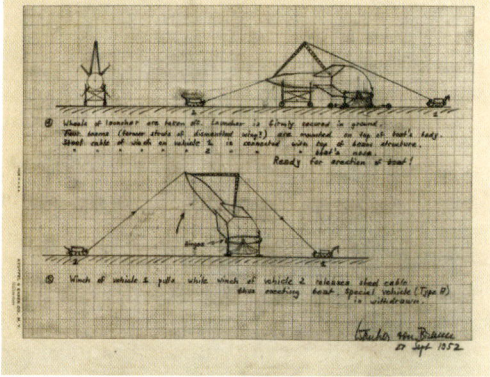

Wernher von Braun, diagram illustrating the
seven steps of rocket-launch preparation,
1952.

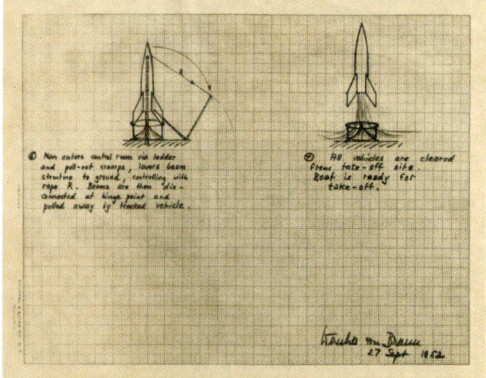

NASA, launch of rocket Bumper 2, 1950. The Bumper rockets were created by NASA using
hardware and calculations retrieved from Germany in the immediate aftermath of WWII.

As both a Nazi rocket engineer and the
father of the United States' space programme,
Wernher von Braun is a compelling and
controversial figure in the history of
spaceflight. Fascinated with astronomy
and rocketry from boyhood, von Braun worked
as the technical director of the Nazis'
V-2 weapon programme. Near the end of the
Second World War, he and key members of his
team were spirited away to the United States
to develop rocket hardware in the desert
of New Mexico - eventually resulting in the
Saturn V rocket, which launched humankind
to the Moon.

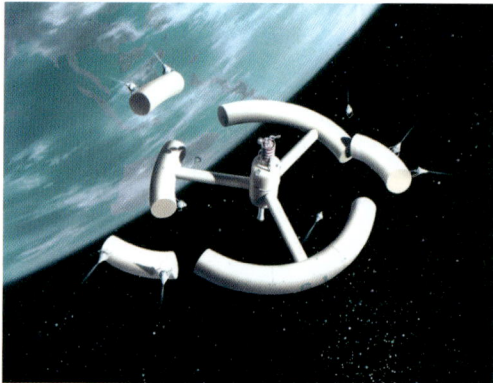

Man in Space, directed by Ward Kimball, film
stills, 1955. Von Braun and other scientists
worked with Disney to create the film
in the hope of increasing public interest
in the space program, © 1955 Disney.

Wernher von Braun at his desk, 1964.

Wernher von Braun by the F-1 engines of the
Saturn V launch vehicle, 1959.

Chesley Bonestell, cover of *Collier's*
magazine, 1952.

Chesley Bonestell, cover of *Collier's*
magazine, 1954. A special issue in honour
of von Braun's plan for a ten-ship
flotilla to Mars.

Originally trained as an architect, Chesley
Bonestell created what is widely thought
to be the most influential astronomical
landscape of all time, *Saturn as seen from
Titan* (1944). Nicknamed 'the painting that
launched a thousand careers', the image
inspired an entire generation of aeronautical
engineers and scientists. It also led to
Bonestell being commissioned to illustrate
Wernher von Braun's influential *Collier's*
series 'Man Will Conquer Space Soon!', an
important cultural marker in the Space Race.

Willy Ley and Wernher von Braun, *Project
Mars*, book-cover artwork designed by Chesley
Bonestell, 1961.

Chesley Bonestell, *Landing on Mars*,
c. 1954-5.

Chesley Bonestell, *Pulling rockets into
upright position for takeoff*, c. 1956.

Chesley Bonestell, *Assembling the ships for
the Mars expedition*, c. 1956.

Chesley Bonestell, *Dust storm on Mars*, 1956.

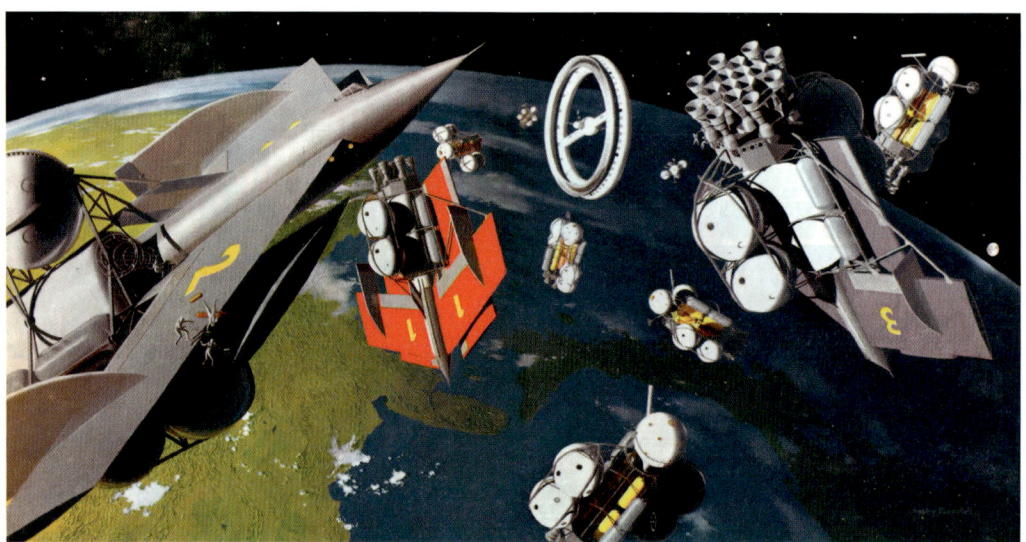

Chesley Bonestell, *Assembling the Mars expedition*, 1953.

Chesley Bonestell, *620 miles above Mars*, 1953.

Chesley Bonestell, *Saturn as seen from Titan*, detail, 1952.

Chesley Bonestell, illustration for Willy Ley's and Wernher von Braun's book *The Exploration of Mars*, 1953.

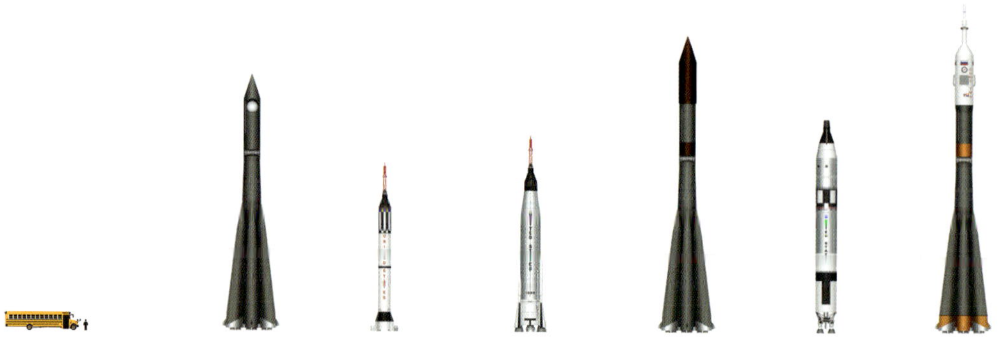

Rocket Name Years Active	VOSTOK 1958-1991	REDSTONE 1960-1961	ATLAS LV-3B 1960-1963	VOSKHOD 1963-1976	TITAN II 1964-1966	SOYUZ 1965-Present
Nozzle Configuration						
Number of Manned Missions	6	2	4	2	10	141
Number of Humans Launched	6	2	4	5	20	269
Payload to Low Earth Orbit	4752kg	Suborbital	1360kg	5900kg	3100kg	7100kg

Tyler Skrabek, *The Rockets of Human Spaceflight*, 2019. An infographic showing how rockets have evolved over the past sixty years.

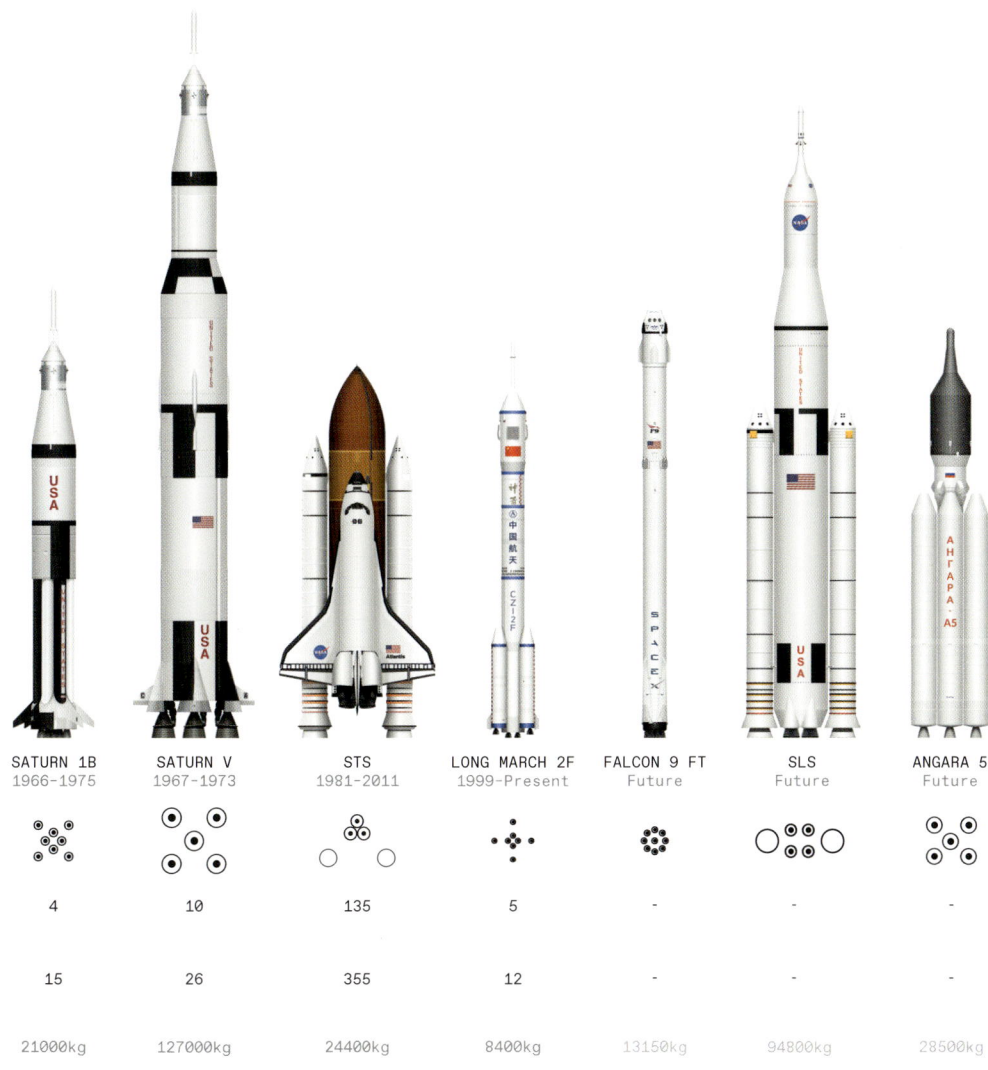

SATURN 1B	SATURN V	STS	LONG MARCH 2F	FALCON 9 FT	SLS	ANGARA 5
1966-1975	1967-1973	1981-2011	1999-Present	Future	Future	Future
4	10	135	5	-	-	-
15	26	355	12	-	-	-
21000kg	127000kg	24400kg	8400kg	13150kg	94800kg	28500kg

NASA, *Orion* spacecraft, render, 2013.

NASA, *Orion* crew module, render, 2007.

The Space Launch System (SLS) is a super heavy-lift expendable launch vehicle currently in development by NASA. It is being designed to replace the retired space shuttle and is a key component of the agency's plans for deep-space exploration.

Alongside the SLS, NASA and ESA are also developing the *Orion* Multi-Purpose Crew Vehicle. The spacecraft is designed to carry four astronauts to low Earth orbit and beyond, and is intended to be the main crew vehicle for potential flights to Mars.

NASA, the Space Launch System's Block 1 rocket booster with the *Orion* spacecraft on its
launch pad, render, 2019.

Lockheed Martin, Mars Base Camp, render, 2017. A NASA-commissioned concept for a crewed laboratory that would orbit Mars and act as a 'base' from which to research the planet.

Lockheed Martin, Mars Base Camp lander, render, 2017. Landing vehicle to carry astronauts between Mars and the orbiting base camp.

SpaceX, Merlin Vacuum engine, 2017. The Merlin engines were developed by SpaceX for its Falcon and Falcon Heavy rockets.

SpaceX, launch of the Falcon 9 rocket on July 14, 2014.

SpaceX, the Dragon spacecraft approaches the ISS on 17 April 2015.

SpaceX, Dragon capsule interior, designed to carry cargo and astronauts to the ISS and other destinations, 2015.

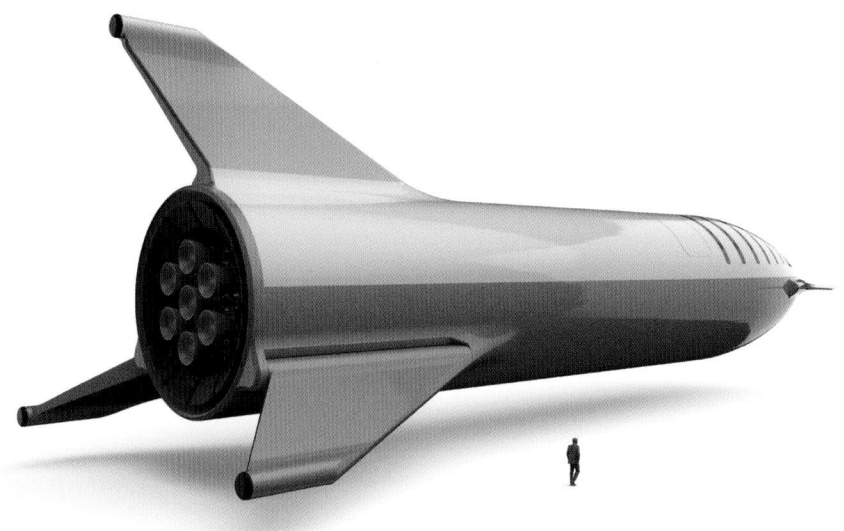

SpaceX, Big Falcon Rocket (BFR) as seen from the side, render, 2018. The BFR was renamed
Starship in November 2018.

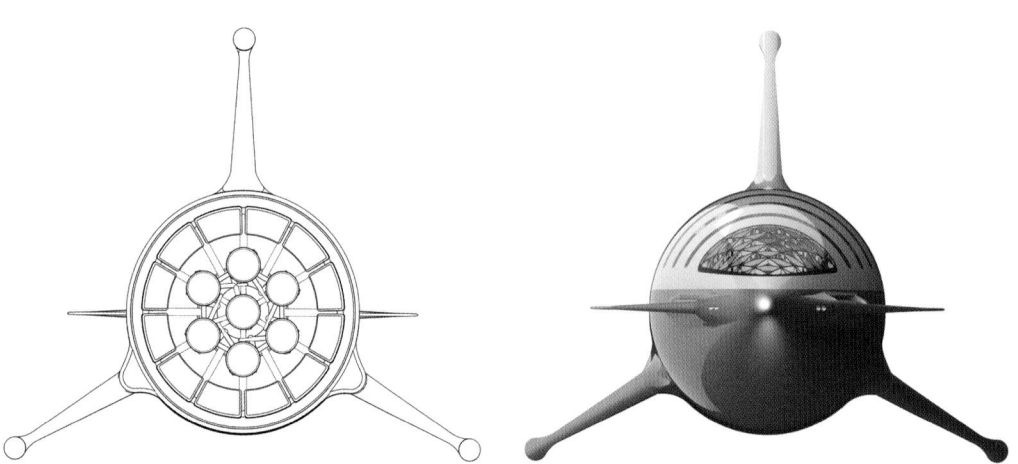

SpaceX, diagram of the rear of the
Big Falcon Rocket, 2018.

SpaceX, Big Falcon Rocket as seen from the
front, render, 2018.

SpaceX, Starship's launch, render, 2018.

SpaceX, separation of the Super Heavy booster and Starship once out of Earth's atmosphere, render, 2018.

Founded by tech entrepreneur Elon Musk, SpaceX is a commercial spaceflight company launched in 2002. It aims to slash the cost of space exploration by applying commercial engineering principles to rocket design, as demonstrated by its pioneering use of reusable launch systems. Musk is committed to human exploration and settlement on Mars, viewing the planet as a viable back-up option should Earth be faced with an 'extinction level event'.

SpaceX, Starship landing on Mars, render, 2018.

SARTORIAL THOUGHTS ON TAILORING SPACESUITS
Anna Talvi

The word 'cyborg' first appeared in 1960 in the *Journal of Astronautics*. The article 'Cyborgs and Space' argued that 'altering man's bodily functions to meet the requirements of extra-terrestrial environments would be more logical than providing an earthly environment for him in space.' One would be free to explore without having to breathe or worry about radiation or muscle atrophy.[1]

Neither the first men and women in space nor their successors were bionic, but the same systems-engineering approach that guided the development of NASA's hardware and software was applied to the conditioning of humans to spaceflight. But a human being is not a machine, and people are more than a system of inputs and outputs.

How should we design for human spaceflight in 2019? As the length of time spent in microgravity increases, human responses to those conditions come into focus. It is not only physiological problems that are amplified, but also – and more importantly – psychological. The role of design-thinkers, and specifically design thinker-doers, is therefore essential for creating new ways of navigating the complex challenges of living in space.

The International Space Station (ISS) has been inhabited continuously since the year 2000. It was designed as a space laboratory,

and the human experience aboard was subordinate to this primary concern.[2] The first *NASA Systems Engineering Handbook* contains only fifteen instances of the word 'human'.[3] Over time, there has been a growing recognition of the importance of human-centred design – a revised version of the handbook, published twelve years later in 2007, acknowledges 'human systems' as one of three key elements next to hardware and software.[4] While this shift is positive, it has yet to manifest itself in the daily lives and daily wear of astronauts.

While NASA's systems engineering results in magnificent rockets, cutting-edge science and technology, and safer space travel, design thinking ensures that we optimise the human experience. It is essential to include designers and design thinking at the start of the problem-solving process, rather than introducing them at the end to make something look nice and user-friendly. Designers are experts in mediating complexity, including that of bringing technology and human aspects together to work seamlessly.

Design thinking is the art of asking better questions. It is about asking 'whys' before, during and after the 'hows'. It is about iterating and real-life testing. Only after n+7 prototypes do we get back to the infant idea, the one that is simple and complex at the same time.

A spacesuit is like an exoskeleton that protects against the hazards of space, just as a tailored suit protects us against earthly hazards like a lack of confidence or a sexist boss. In both cases, greater protection brings with it greater restriction.

ILC Dover, A-7L Extravehicular Mobility Unit worn by Buzz Aldrin on the moon, 1969.

Originally, the preferred option for the Apollo suit was a hard-shell masterpiece of engineering. These sci-fi-looking AX and RX suits – based on the concept of the US Air Force vacuum chamber – were difficult to manoeuvre in. The first suits to walk on the Moon were, in contrast, soft, white and stitched together on a sewing machine by seamstresses of ILC Dover – a division of the company better known as Playtex, famous for manufacturing women's bras and girdles. These A7L spacesuits worked comfortably on the human body and not just the engineer's calculation sheet.[5] To prove this to NASA, ILC Dover presented the spacesuits on men playing football.

In fact, spacesuits have a lot in common with bespoke tailoring. First, each suit is custom-made for a specific person. Second, small subtleties that are hidden beneath the lining are what make a good suit. The creation of the suit is the result of decades of accumulated knowledge and refinement of technique. It was this tacit knowledge in the hands of the seamstresses that brought the Apollo suits into being.

While extensive research and development has been done on EVA (Extravehicular Activity) and flight suits using the systems-engineering approach, very little research has addressed astronauts' daily wear for living on space stations like the ISS, or on longer journeys, such as the one to Mars. It is especially important to address this issue from a human-centred design and sartorial perspective, one that addresses human physiology and psychology not just through inputs and outputs, but with all its internal complexities.

This is necessary because the length of time spent in microgravity is increasing greatly. The first (wo)manned round trip to Mars scheduled for the 2030s will take about three years, and the difficulties of staying fit and healthy for such an extended period of time in space are considerable. It is likely that the first mission to Mars will be all-female. Studies have shown that female astronauts cope better with the psychological stresses of working in a small space with a limited crew; they also require fifty per cent fewer calories, which creates savings in fuel and money,[6] and a single-sex crew lowers the risk of astronauts getting frisky.[7] The (wo)man as machine is back on the table. Fifty years after the first moon leap, the first all-female spacewalk was cancelled because of a scarcity of spacesuits that fitted the female form.[8] Going forward, we need better designs: ones that will keep astronauts – both male and female – healthy, safe and comfortable.

Just as the absence of gravity changes the *modus operandi* of the human body, we also need to rethink the way we design for life without gravity. Conventional garment construction is built on the principle of gravity resting the suit on the wearer's shoulders. Seams traditionally run vertically, falling in line with the direction of the Earth's gravity. Once we are released from gravity, shouldn't we also release our clothes from static block patterns?

The microgravity-wear I have developed uses swirling seams that wrap the garment around the floating body and rest it entirely around the neutral body posture. Instead of using standard 2D block patterns, these garments are constructed directly on physical, moving bodies, using kinetic garment construction. This results in improved fit, adaptability to a wide range of movements, including those specific to life in a spacecraft, and helps to liberate micro-gravity garments from standard static, straight legs and restrictive armholes. Using 3D body scanning and virtual 3D technology helps to simulate how the body's proportions will shift in microgravity and to fit the garments for those expected measurements.[9]

The biggest problems that face bodies relieved of gravity is a loss of bone density and muscle atrophy.[10]

Anna Talvi, Antagonist Exo. Muscle bodysuit and X.Earth olfactory gloves-mask, 2018.

To counteract this, astronauts do two hours of exercise a day, with their bodies restrained by bungee cords. This would not be enough to keep the bone density and muscle mass at safe levels on a three-year round trip to Mars. Mechanical counterpressure suits that squeeze the body from feet to shoulders have been tested on the ISS, but lack comfort and ease of use. An alternative, electromagnetic muscle stimulation has also been tested on astronauts, but the 150 volts running through your tissue is invasive and unpleasant, as well as altering the cellular make-up of muscles, blood vessels and nerves.

If we approach this problem in a designerly way, then we can stop our bodies from diminishing without also reducing our physical and mental well-being. The Antagonist Exo.Muscle bodysuit we are developing uses

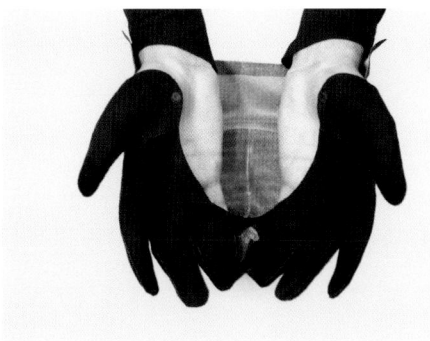

smart membranes for engaging the postural muscles, which are normally relaxed in microgravity. This 'wearable gym' also helps to load the weight-bearing bones in a more natural, non-invasive way. A fully functioning solution that is effective, compatible with daily life and ready to use is still in development. But who better to tackle these problems than those with a sartorial eye, in collaboration with doctors, sports physiologists and materials scientists?

I have designed this optimal set of garments for a prolonged stay in microgravity environments, considering the vast array of other challenges that come with life at zero gravity. These garments are tailored to the terrestrial body in celestial space. They give the wearer comfort and confidence, while helping them stay fit and healthy.

Anna Talvi, X.Earth olfactory gloves-mask, 2018.

Among the challenges that a designer must address are extremely limited cargo, inability to wash clothes and the fact that sweat does not evaporate like it does on Earth, but instead creates an aura around your body.[11] The textiles and membranes used in my microgravity-wear are therefore very lightweight, hyper-wicking, anti-microbial and easy-care. To combat mental challenges – such as a longing for normal sleep, space-anxiety and the knowledge that it is impossible to turn back home – I have created the X.Earth olfactory gloves-mask. The X.Earth can conjure memories of one's home through smell. They are custom-made for each individual and contain a personal 'Earth-memory smell-scape capsule'. When the wearer puts the gloves to their face and inhales, they can be reminded of anything from the scent of someone they love to the smell of rain. These familiar smells trigger the limbic system – part of the brain most susceptible to radiation – bringing calm and mental well-being to the long journey.

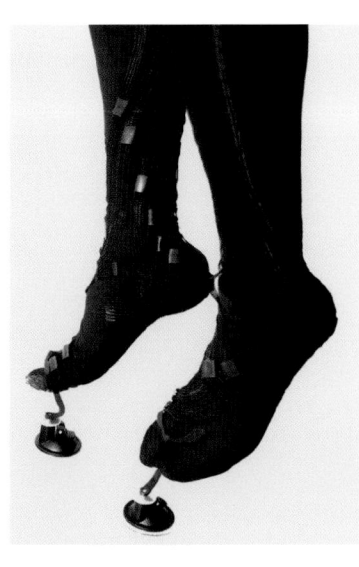

Finally, my microgravity-wear set provides a less bothersome method of anchoring oneself in this floating world,[12] including spaceplugs – handy toe sockets that allow astronauts to fix themselves onto the spaceship's surfaces. The microgravity-wear is made to take advantage of newly found perks of life without gravity, such as the fulfilment of every child's dream to be able to jump into two trouser legs at once, or not to need to wear shoes – after all, you won't be walking.

Design thinking is the key to understanding the problems and opportunities of human spaceflight. We are still not ready to go to Mars as cyborgs. We are just about ready to go as humans.

Anna Talvi, Spaceplugs and
Antagonist Exo.Muscle bodysuit,
2018.

1   Manfred E Clynes and Nathan S Kline,
    'Cyborgs and Space', in *Journal of
    Astronautics* (September 1960), http://
    www.guicolandia.net/files/expense/
    Cyborgs_Space.pdf [Accessed 11 February
    2019].

2   Tibor Balint, 'Design Space for Space
    Design: Humanly {S:pace} Constructs
    across Perceptual Boundaries', PhD thesis
    (Royal College of Art, 2016).

3   National Aeronautics and Space
    Administration, *NASA Systems Engineering
    Handbook* (1995), https://www.isibang.ac.
    in/~library/onlinerz/resources/
    NASAengineeringhandbookpdf.pdf
    [Accessed 11 January 2019].

4   National Aeronautics and Space
    Administration, *NASA Systems Engineering
    Handbook* (2007), https://www.nasa.gov/
    connect/ebooks/nasa-systems-engineering-
    handbook [Accessed 11 January 2019].

5   Nicholas de Monchaux, *Spacesuit:
    Fashioning Apollo* (Cambridge, MA:
    MIT Press, 2011).

6   Kate Greene, 'An All-Female Mission
    to Mars: as a NASA guinea pig, I
    verified that women would be cheaper
    to launch than men', in *Slate Magazine*
    (19 October 2014), https://slate.com/
    technology/2014/10/manned-mission-to-
    mars-female-astronauts-are-cheaper-to-
    launch-into-outer-space.html
    [Accessed 1 July 2019].

7   'Mars missions may be all-female to avoid
    astronauts having sex during 1.5-year
    journey', in *The Express Tribune* (29
    September 2017), https://tribune.com.pk/
    story/1519454/mars-missions-may-female-
    avoid-astronauts-sex-1-5-year-journey/
    [Accessed 1 July 2019].

8   Thalia Patrinos, 'Spacewalk
    Reassignments: what's the deal?',
    *National Aeronautics and Space
    Administration* (27 March 2019),
    https://www.nasa.gov/feature/spacewalk-
    reassignments-what-s-the-deal
    [Accessed 1 July 2019].

9   Limbs get smaller and the torso gets
    bigger in microgravity, due to blood
    redistribution.

10 The rate of bone loss in microgravity
    is approximately ten times that of
    osteoporosis on Earth, and the loss of
    muscle mass in one week in microgravity
    can be up to twenty per cent higher.

11 This is because there is no convection
    in zero gravity. Having an aura of
    unevaporated sweat around the body
    leads to skin problems.

12 The current ISS astronauts develop
    painful 'lizard skin feet' and bone
    deformations on their feet from strapping
    their feet behind wall-mounted elastic
    straps when they need to work in a
    static position without floating away.

Wiley Post with Russell Colley of BF
Goodrich, the world's first practical
pressure suit for use at high altitude, 1934.

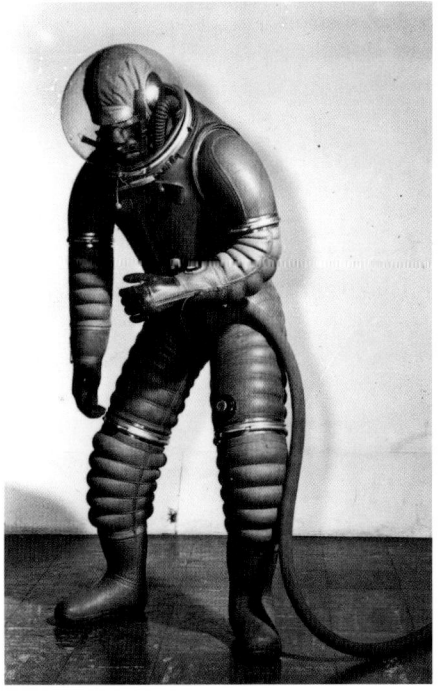

Russell Colley of BF Goodrich, Goodrich
XH-5, commonly known as the 'tomato worm
suit', c. 1943-50.

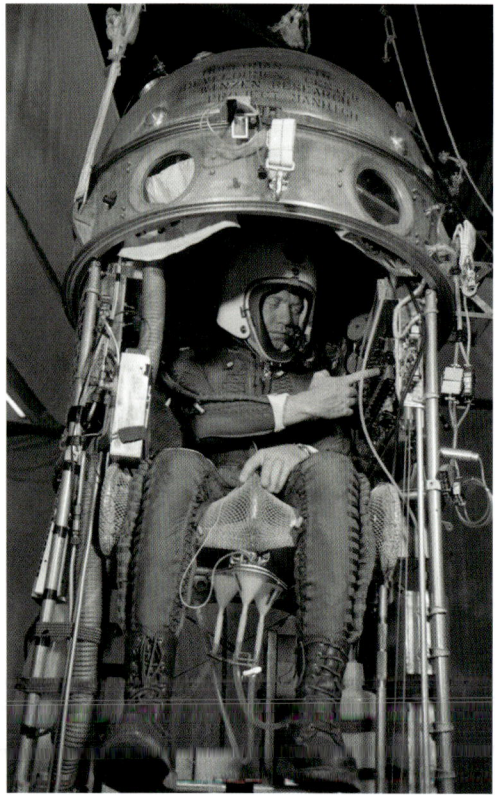

Joseph Kittinger's record-breaking jump
from the helium-balloon-supported Excelsior
gondola, 1960.

Joseph Kittinger in a helmet and pressure
suit, preparing for his parachute jump
from a helium-balloon-supported gondola
at stratospheric height, 1957.

Colonel Joseph Kittinger set a world
record in 1960 for the highest skydive,
free falling from his stratospheric balloon
31 kilometres (19.3 miles) above the Earth.
The jump was part of Project Excelsior,
a military operation designed to study
parachute systems at high altitude.

Kittinger wore a modified David Clark
MC-3A partial pressure suit during the
jump, with additional layers of clothing
to protect him from the extreme cold.
During the mission he became one of the
first people to ever see the curvature
of the Earth.

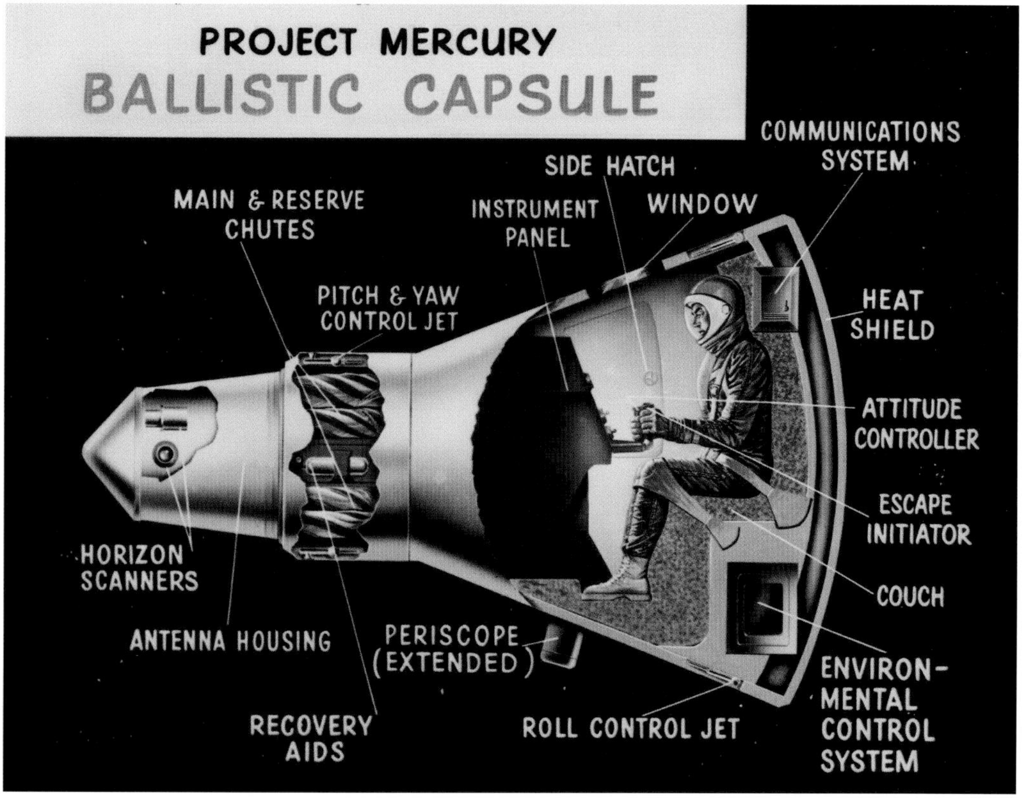

NASA, cutaway diagram of the *Mercury* space capsule, 1960.

BF Goodrich, Mercury spacesuit worn by
Alan Shepard in the *Mercury* spacecraft, 1961.

Pressure suits from left to right from top: BF Goodrich, Mark IV, 1950s; BF Goodrich, Mark V, 1968; BF Goodrich, MR-3, 1961; BF Goodrich for Hamilton Standard, XN-20 developmental suit, 1964; ILC Industries, A5-L, 1966; ILC Industries, 'Lobster Shell', 1965;

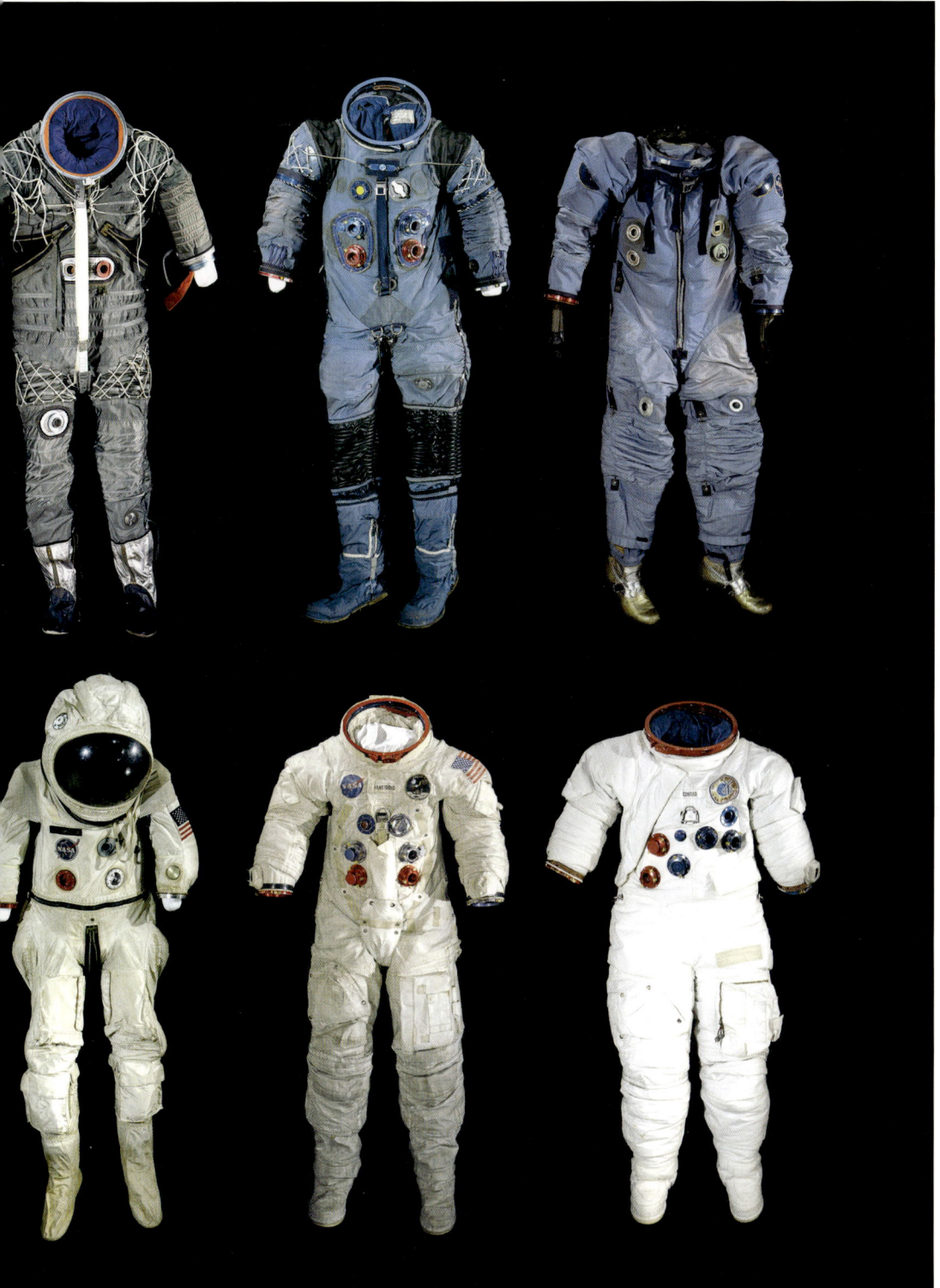

Hamilton Standard, Manned Orbiting Laboratory, 1966-7; Hamilton Standard, MH-7, 1966-7; David Clark Company, A1-C, 1966; David Clark Company, G5-C, 1965; ILC Industries with Hamilton Standard, A7-L, 1969; ILC Industries with Hamilton Standard; A7-LB, 1973.

NASA with Gary L Harris and Pablo de León, North Dakota eXperimental 1 (NDX-1) Mars
spacesuit prototype, 2007. Using modern, lightweight materials, the NDX-1 is meant to be
a self-contained machine that could protect astronauts from the cold Martian atmosphere
and dust storms.

NASA with Gary L Harris and Pablo de León, a test subject performs a soil-collecting experiment wearing the NDX-1 spacesuit at the Lunar Regolith Lab at the Kennedy Space Center, 2015.

A test subject collects a sample while wearing the NDX-1 spacesuit in the Utah desert, 2012.

Prof. Dava Newman (MIT: Inventor, Science and Engineering) with Guillermo Trotti (AIA, Trotti and Associates: Design) and Dainese (Design), BioSuit™, 2013.

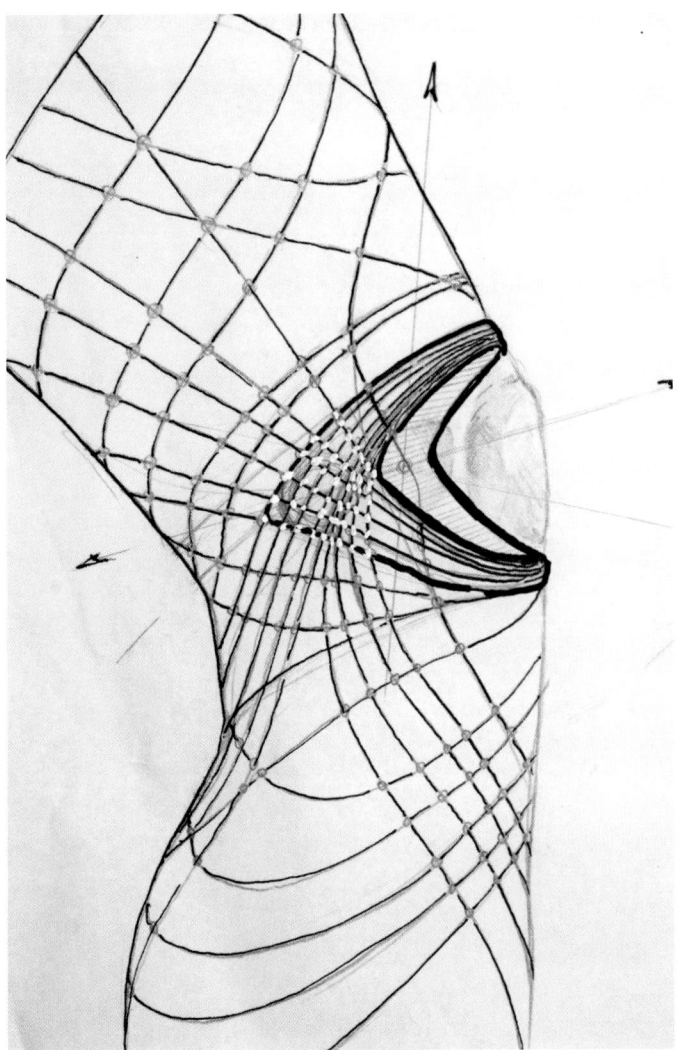

MIT students, sketch of prototype no. 3, c. 2005.
Invented by Prof. Dava Newman (MIT: Inventor, Science
and Engineering). Sketch of the knee of the BioSuit™.

Dava Newman is a leader in spacesuit design, dynamics and control of astronaut motion. Over the course of her career at NASA and MIT, she has developed a revolutionary EVA suit design, called the BioSuit™. The smart-fabric suit provides pressure through compression rather than with pressurized gas. This allows for increased mobility, a key requirement for astronauts on future Mars missions.

SpaceX, spacesuit, 2017. Designed in-house by SpaceX, each custom-tailored suit is meant to provide a pressurised environment for all crew members aboard Dragon in atypical situations such as cabin depressurisation. Features include a 3D-printed spacesuit helmet, touchscreen compatible gloves and a flame-resistant outer layer.

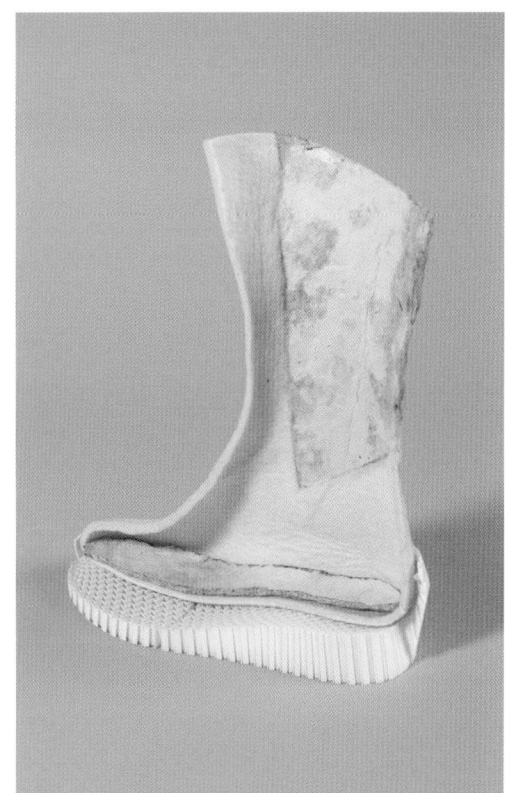

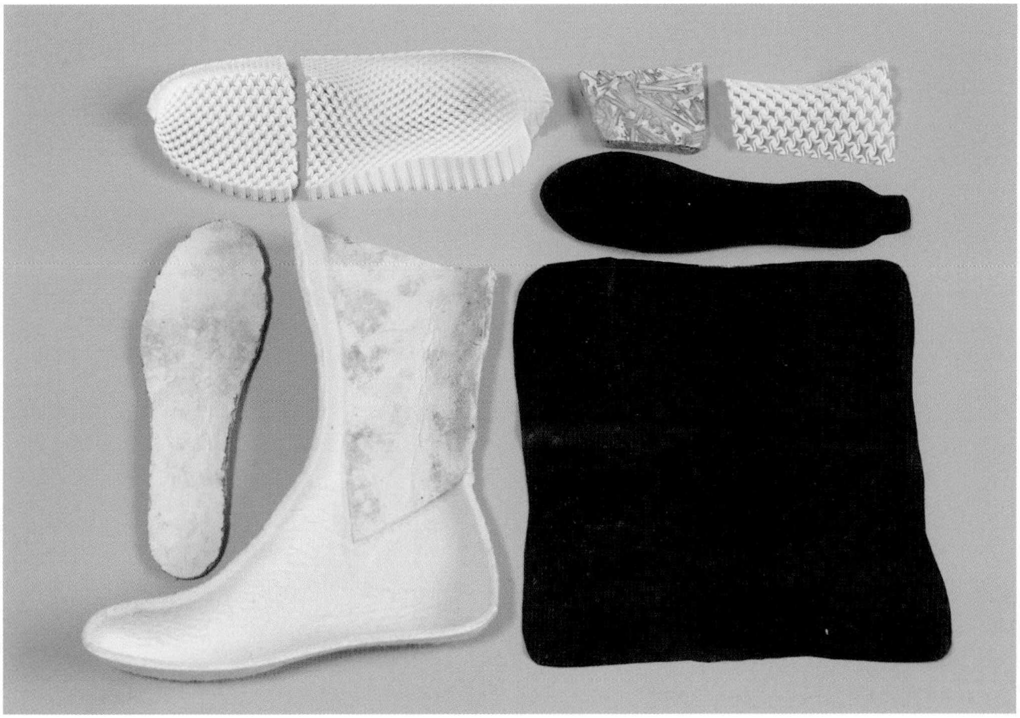

Liz Ciokajlo (OurOwnsKIN) and Maurizio Montalti (Officina Corpuscoli), CASKIA/Growing a Mars Boot, 2018. A prototype Mars boot grown from human sweat and fungal mycelium.

Lucy McRae, 'Astronaut Aerobics', installation at the *Dezeen and MINI Frontiers exhibition: the future of mobility*, London Design Festival, 2014. Drawing inspiration from NASA's lower-body negative-pressure device from the 1960s, this speculative and immersive installation invited visitors to experience being 'vacuum-packed' in preparation for long-distance space travel.

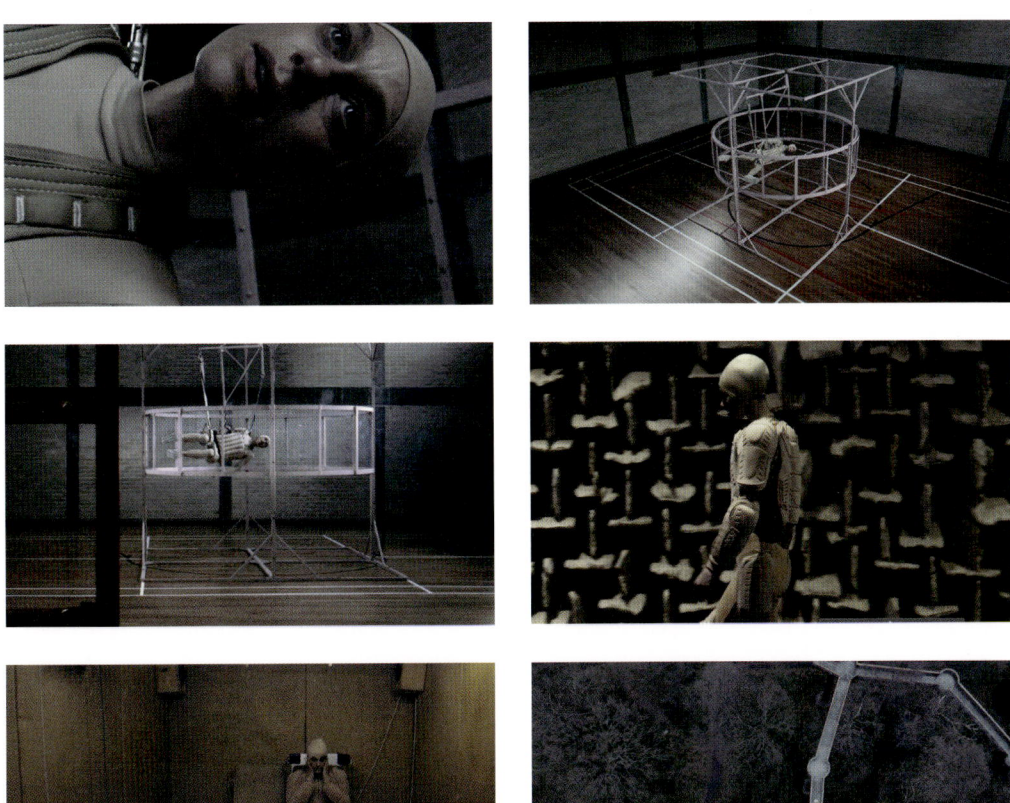

Lucy McRae, *Institute of Isolation*, film stills, 2016.

According to a recent report by NASA, two of the top ten risk factors for a future Mars mission are linked to psychological well-being. The journey to Mars would require astronauts to live for a sustained period of time in a dangerous, confined and isolated environment, pushing the usual boundaries of human endurance. The short film *Institute of Isolation*, by filmmaker and artist Lucy McRae, is a fictional examination of the ways travellers to outer space could use architecture and design to train their bodies for life in extreme isolation. McRae partnered with several designers to create a collection of speculative props for the film, including a microgravity trainer and spacesuit that would help prepare the body for life in space. McRae considers prolonged isolation as an extreme experience, and is interested in questioning how we can design for it.

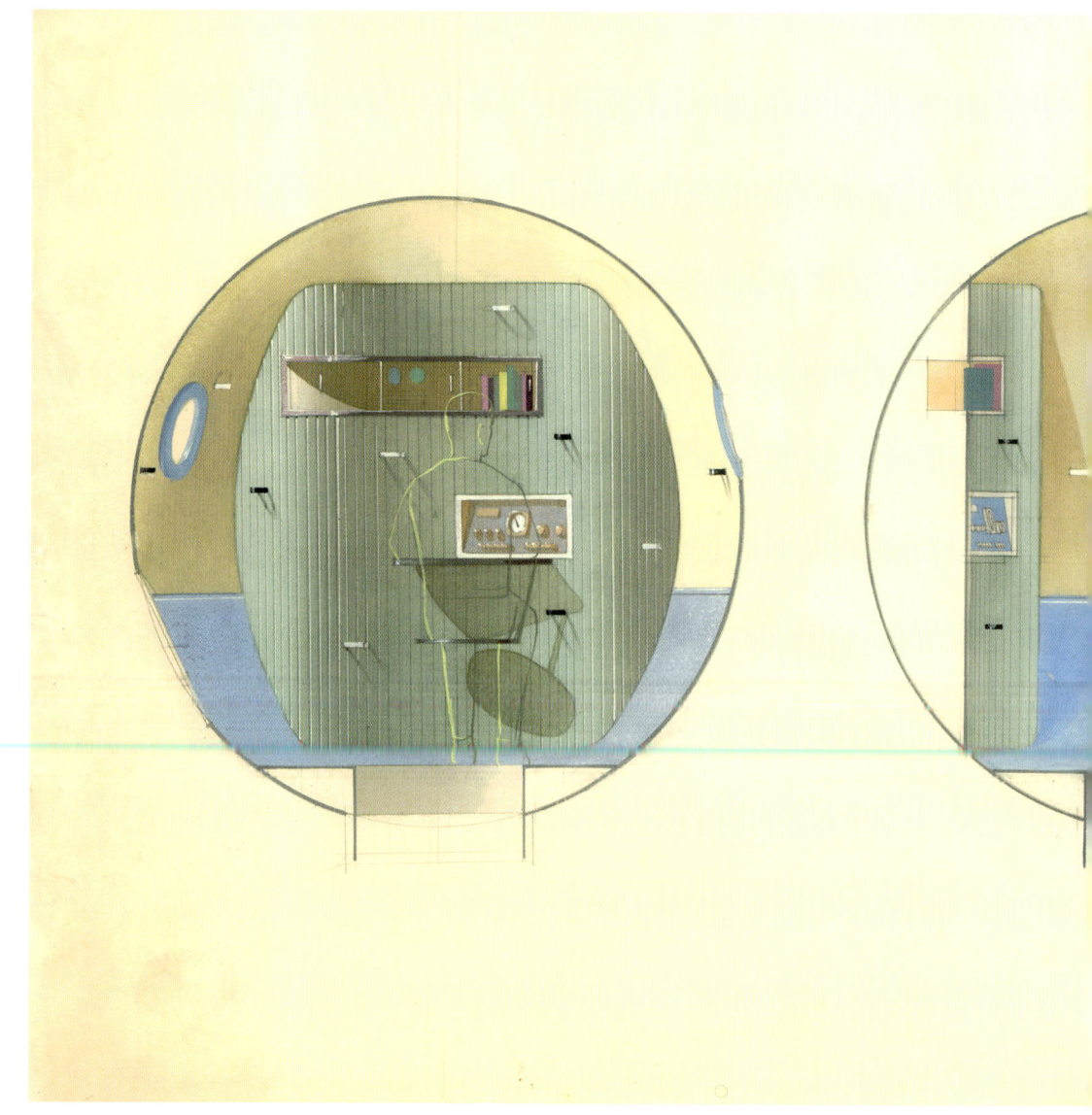

Galina Balashova, initial sketches of the interior of the Lunniy Orbitalny Korabl
(Lunar Orbital Craft, LOK), 1966. The Soyuz 7K-LOK was a Soviet spacecraft designed
for manned lunar orbits.

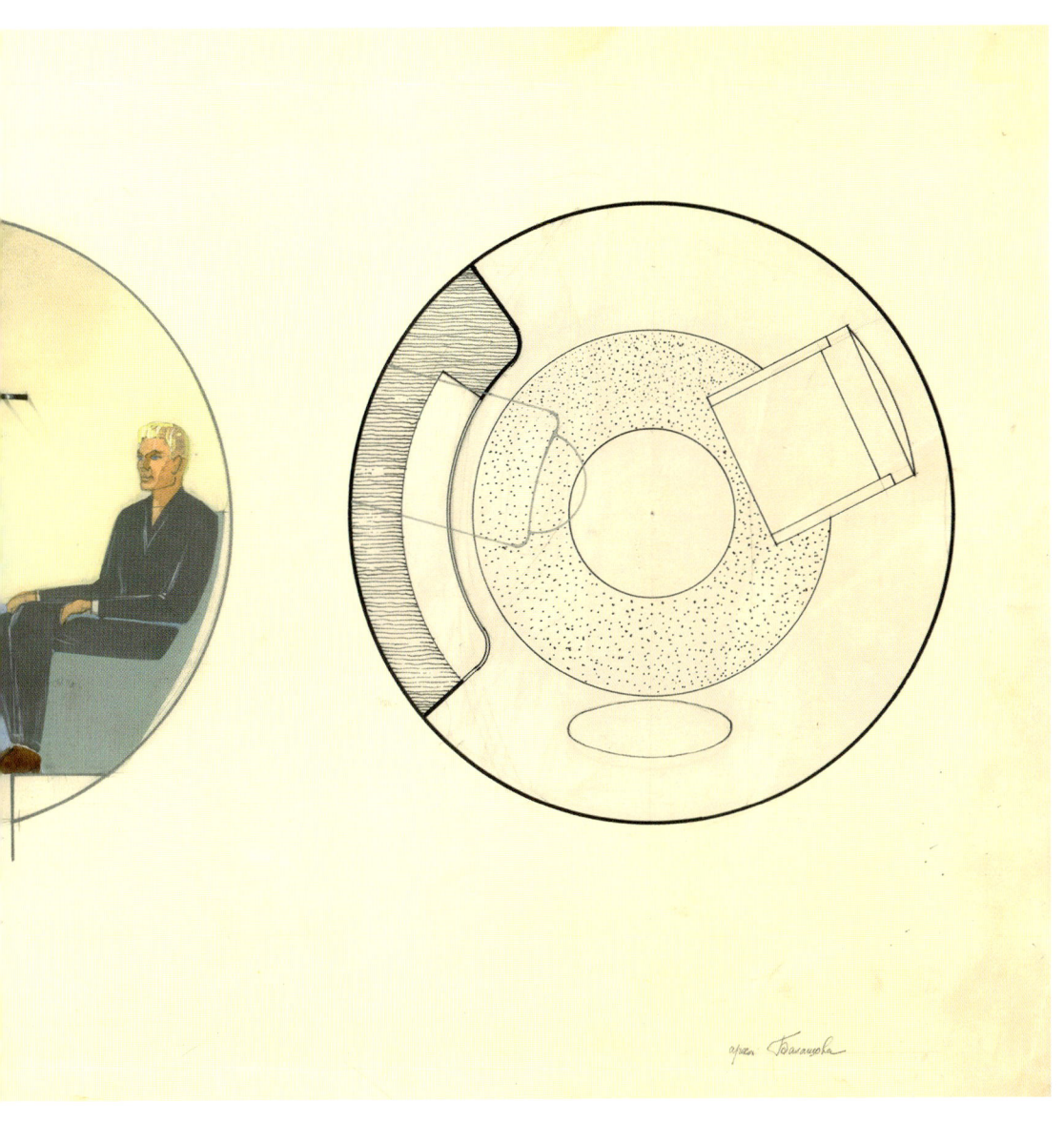

The only architect ever hired by the Soviet space programme, Galina Balashova had considerable influence on the design and aesthetics of Soviet spacecraft. Originally employed to work on buildings for the programme, she was soon asked by Sergei Korolev to design interiors for the Soyuz, Salyut and *Mir* spacecrafts. Her work is still in use on the ISS today.

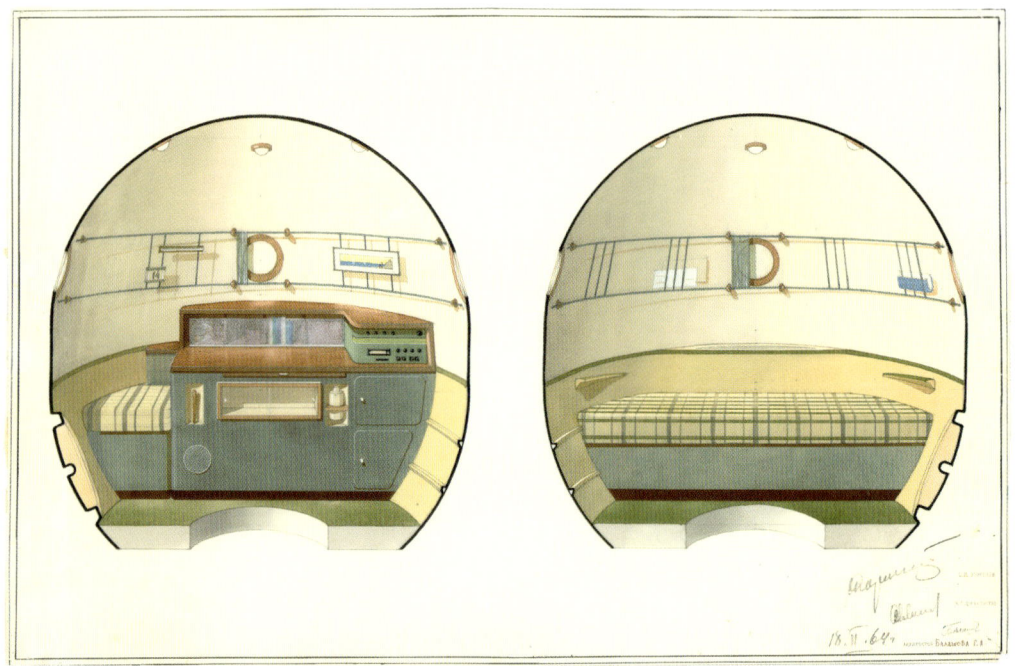

Galina Balashova, final design for the interior of the LOK orbital module, 1964.

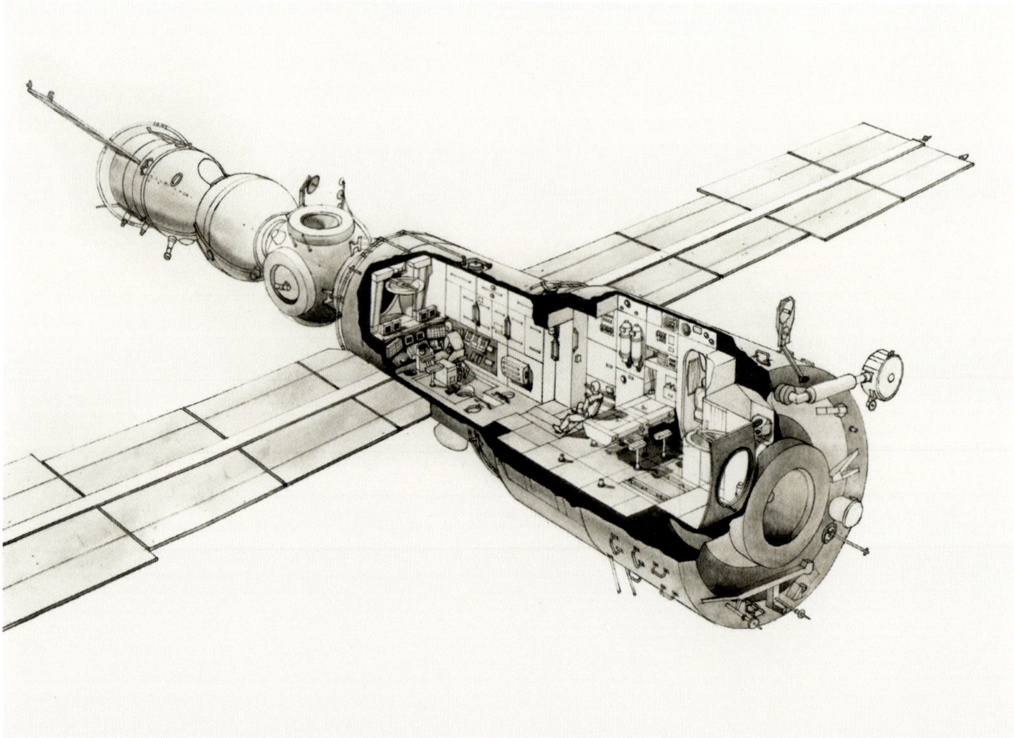

Galina Balashova, *Mir* space station, 1981.

Galina Balashova, design for the workspace of the *Mir* space station, 1980.

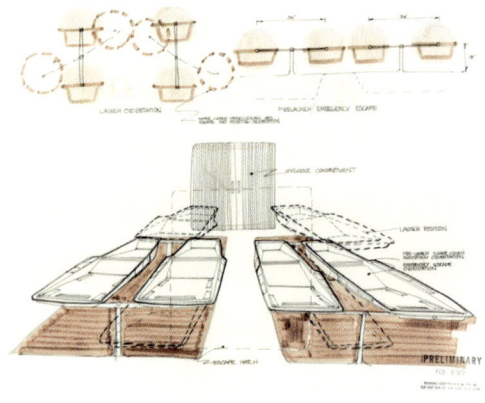

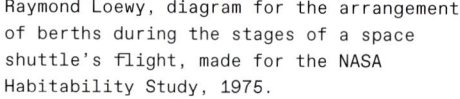

Raymond Loewy, diagram for the arrangement of berths during the stages of a space shuttle's flight, made for the NASA Habitability Study, 1975.

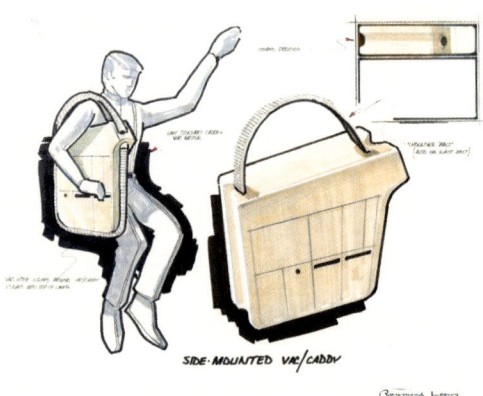

Raymond Loewy, sketch for the NASA Habitability Study, 1969.

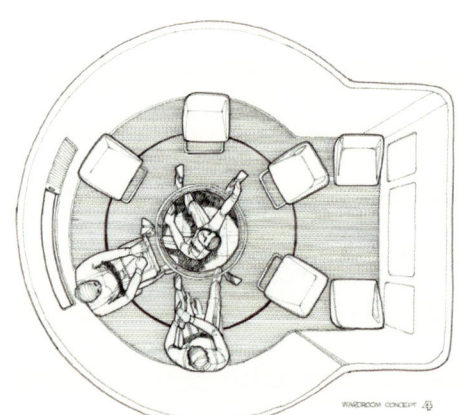

Raymond Loewy, sketch of the aerial view of figures in a wardroom, made for the NASA Habitability Study, 1969-75.

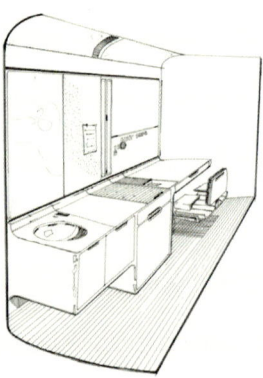

Raymond Loewy, perspective view of a work station made for the NASA Habitability Study, 1969-75.

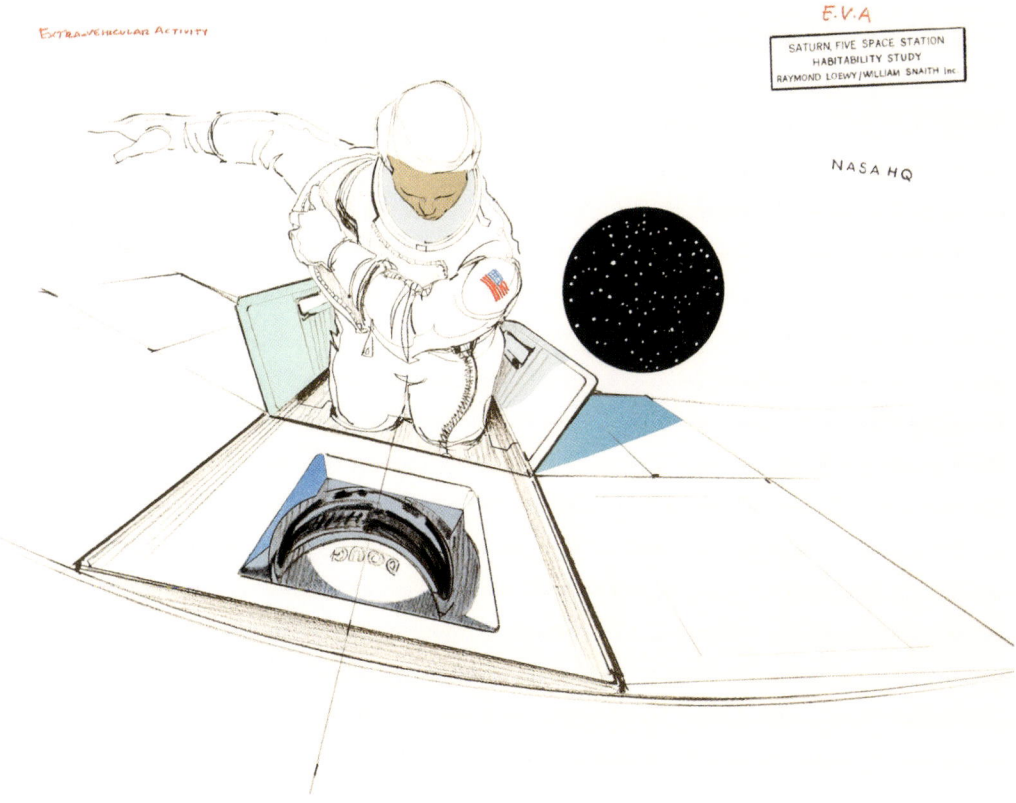

EXTRAVEHICULAR ACTIVITY

E.V.A

SATURN FIVE SPACE STATION
HABITABILITY STUDY
RAYMOND LOEWY / WILLIAM SNAITH Inc

NASA HQ

Raymond Loewy, sketch of a device facilitating the transfer from working overalls to the EVA spacesuit before leaving *Skylab* through a hatch, 1978.

Hailed as 'the father of industrial design', Raymond Loewy was employed by NASA as Habitability Consultant from 1967 to 1973. Tasked with designing the interior of the *Skylab* space station, he has anecdotally been credited as the first person to suggest putting a window on a spacecraft. While many of his suggestions were initially dismissed as superfluous by NASA engineers, his considered treatment of the interior made *Skylab* the most comfortable spacecraft to date.

Raymond Loewy, *Skylab* drawing, 1978.

NASA-S-74-5151

Raymond Loewy, *Skylab* axiometric drawing, 1974.

NASA, members of the first manned *Skylab* mission dining on *Skylab* space food, 1973.

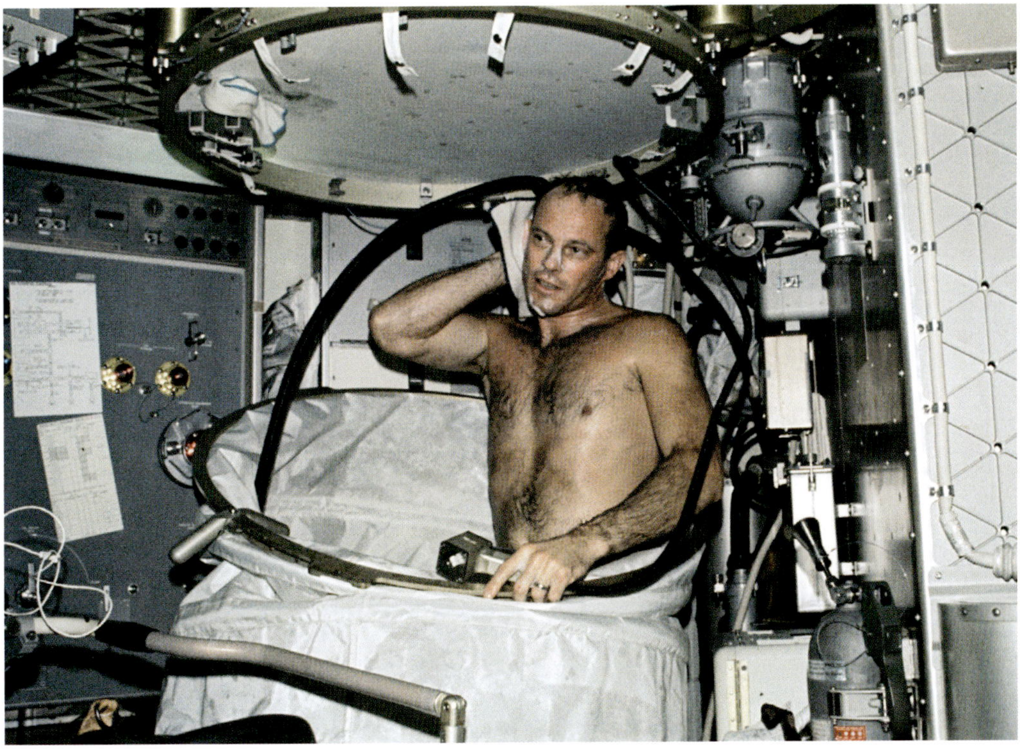

NASA, astronaut Jack R Lousma, bathing in the crew quarters of the *Skylab* space station, 1973.

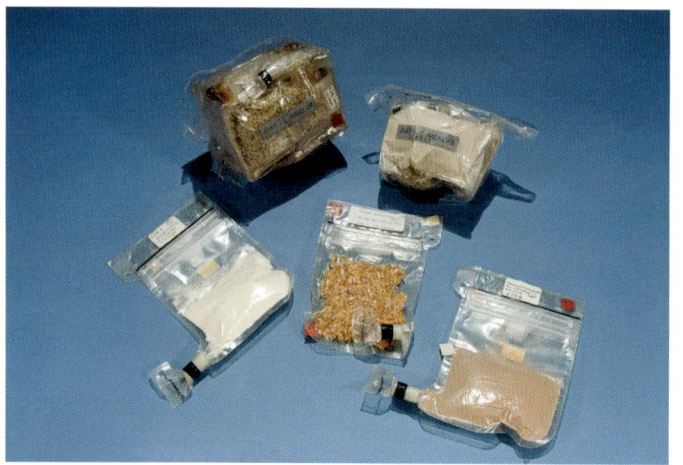

NASA, food for the Apollo missions, 1968-72.

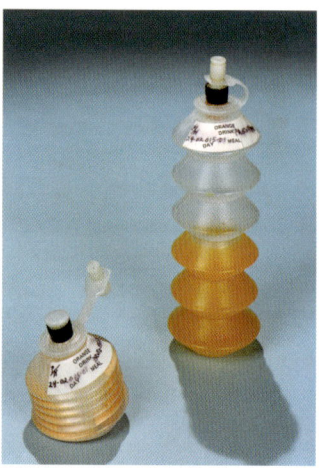

NASA, *Skylab* drink
containers, 1972.

NASA *Skylab* food and tray, 1974.

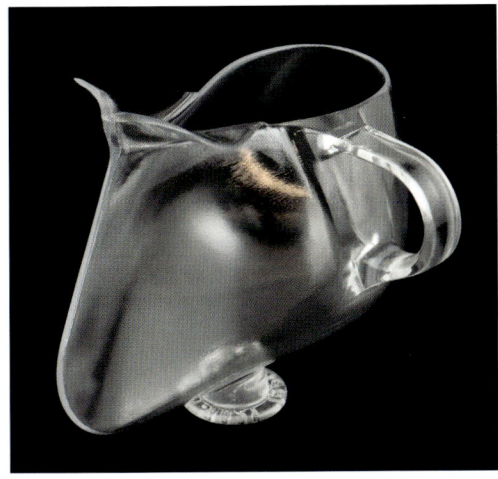

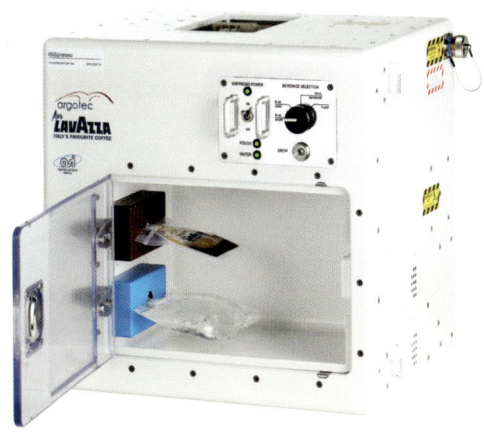

Spaceware, coffee cup for drinking in space, 2016. Coffee cups designed to be used with the ISSpresso machine.

Lavazza, ISSpresso, 2015. An espresso machine designed for use in space.

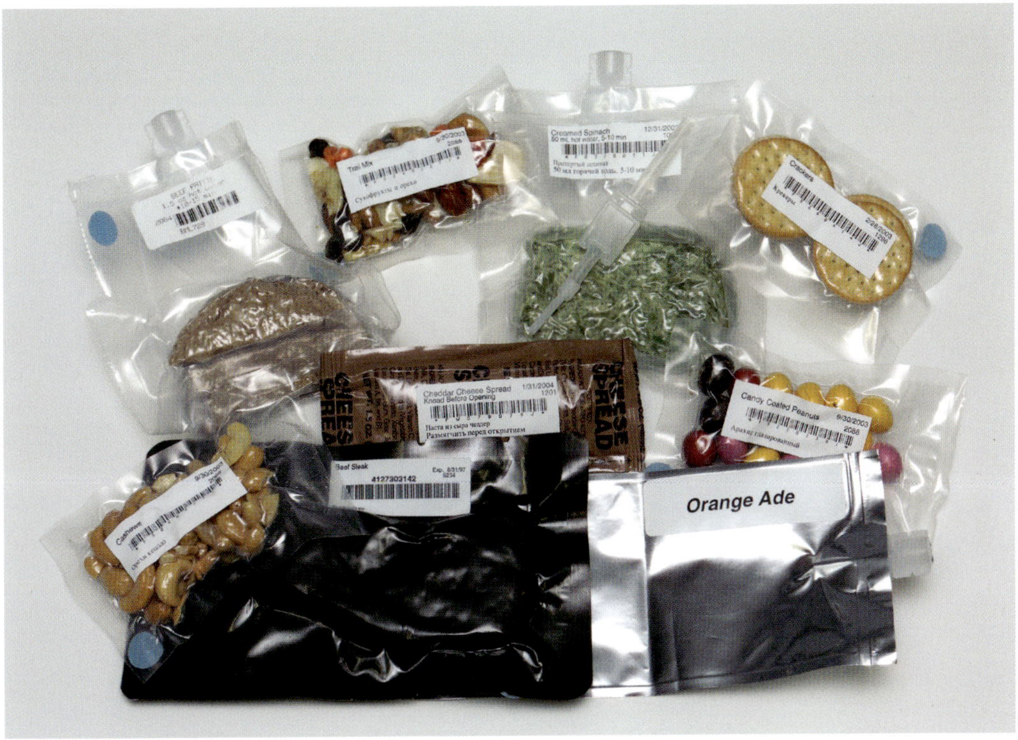

NASA, ISS space food assortment, 2017.

Feeding astronauts in space has always been a challenge. Based on army survival rations, food for the early space programmes consisted of pureed goods packed into aluminium tubes to be sucked through a straw. This later developed into freeze-dried food packages that could be rehydrated on-board. Astronauts on the ISS today can choose from over 100 different food stuffs; however, most still report missing food on Earth, such as fresh fruit and vegetables.

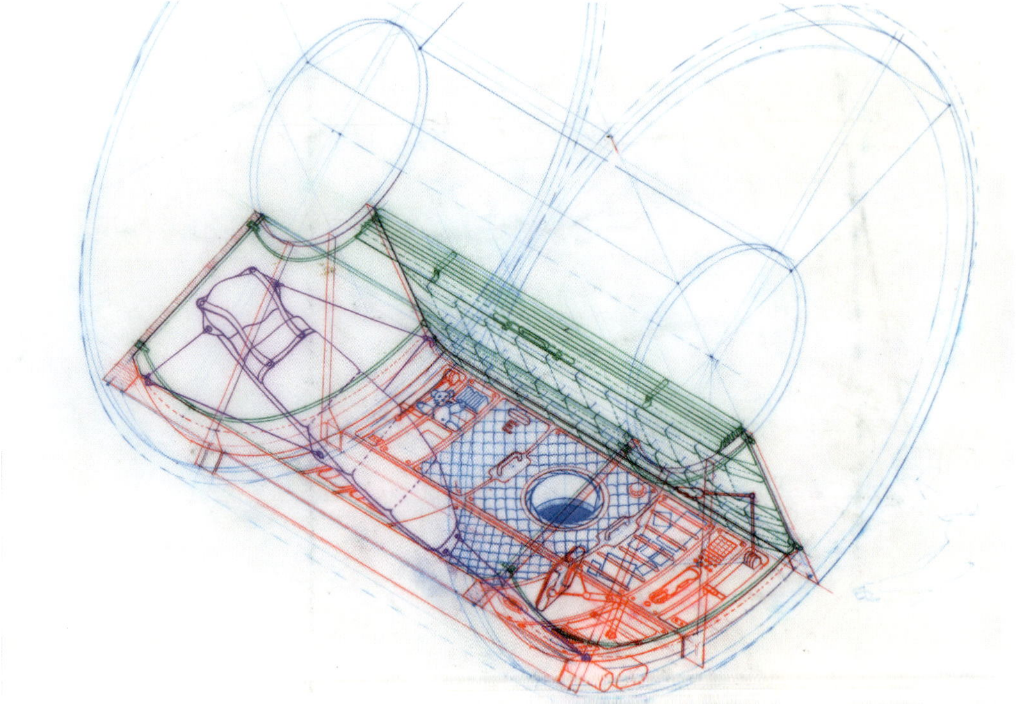

David Nixon, concept drawing of an adjustable radial sleep compartment for the
Space Station (now known as the ISS), 1984.

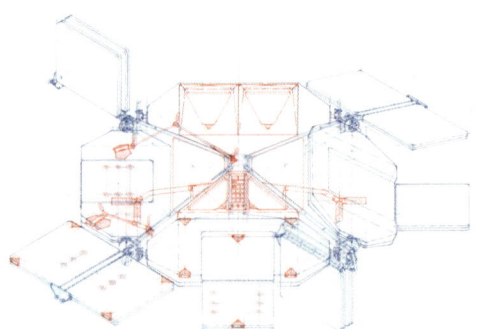

David Nixon, layout drawing of a table
for the habitation module of the Space
Station (now known as the ISS), 1988-9.

David Nixon, working prototype of a
table for the Space Station (now known
as the ISS), 1988-9.

In the 1980s, the Reagan administration
approved plans to start construction
of the International Space Station. With
habitability now a key consideration,
NASA employee Marc Cohen spearheaded an
initiative to bring architects in to work
on the programme, resulting in a number
of radical design proposals from architects
such as David Nixon and Constance Adams.

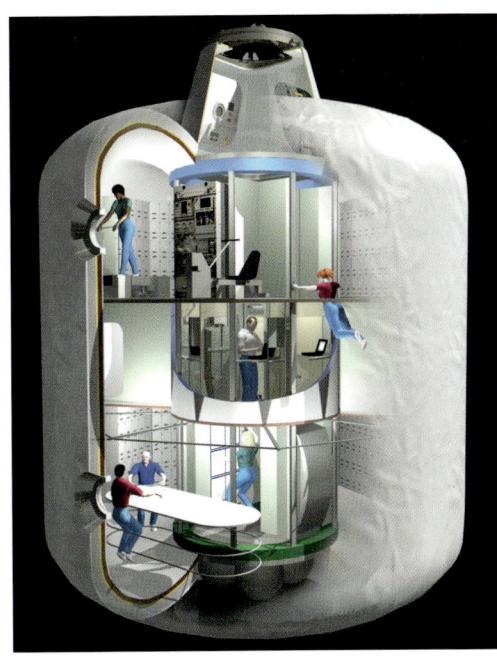

Constance Adams/NASA, Artist rendering of
the Transhab interior workspace attached
to the ISS, Johnson Space Center, Houston,
TX, 2000.

Constance Adams/NASA, Artist rendering of
Transhab, vertical design idea for maximum
usable space, Johnson Space Center, Houston,
TX, 1998.

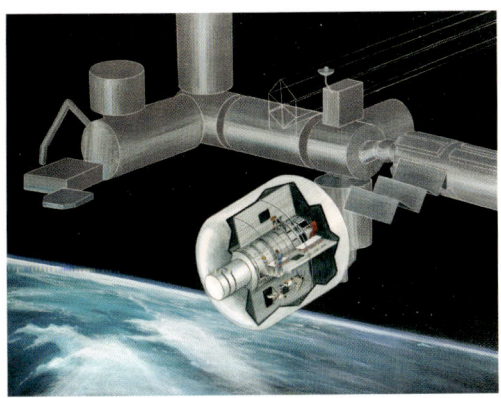

Constance Adams/NASA, TransHab prototype
designed for travel to Mars, concept, 1997.

Constance Adams/NASA, Transhab prototype
and vacuum chamber, Johnson Space Center,
Houston, TX, 1998.

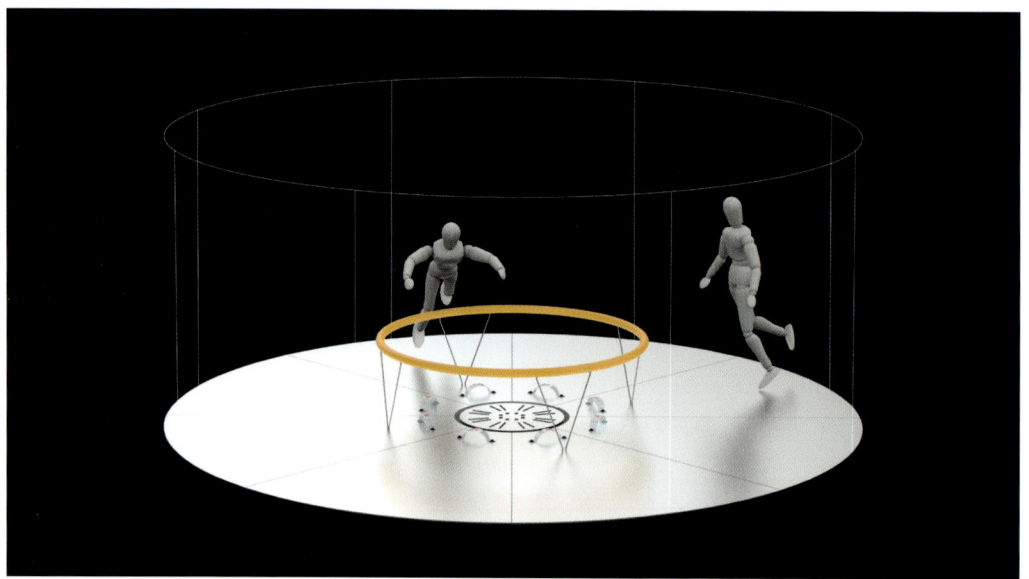

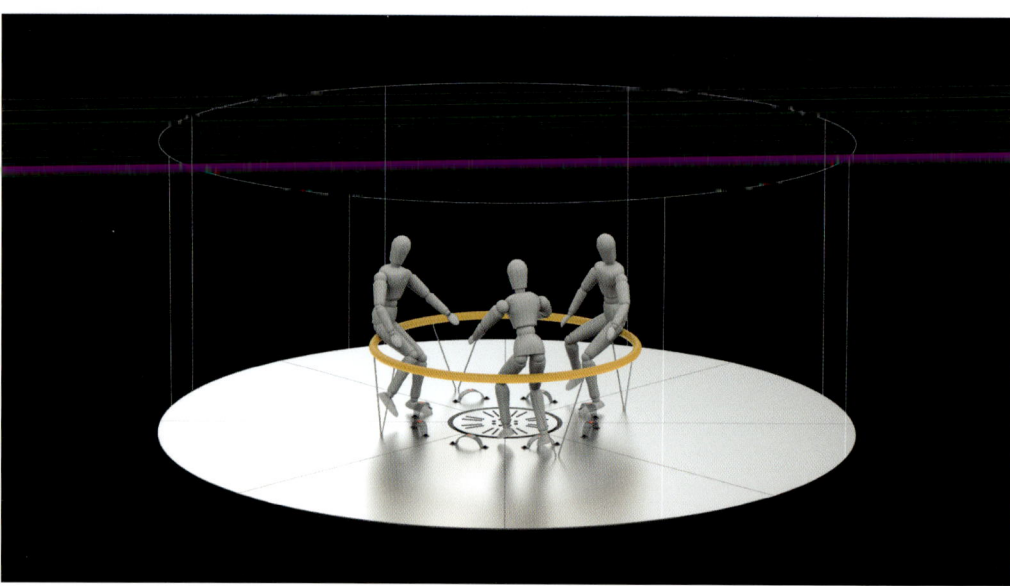

Konstantin Grcic, POW WOW table, commissioned by the Design Museum, 2019.

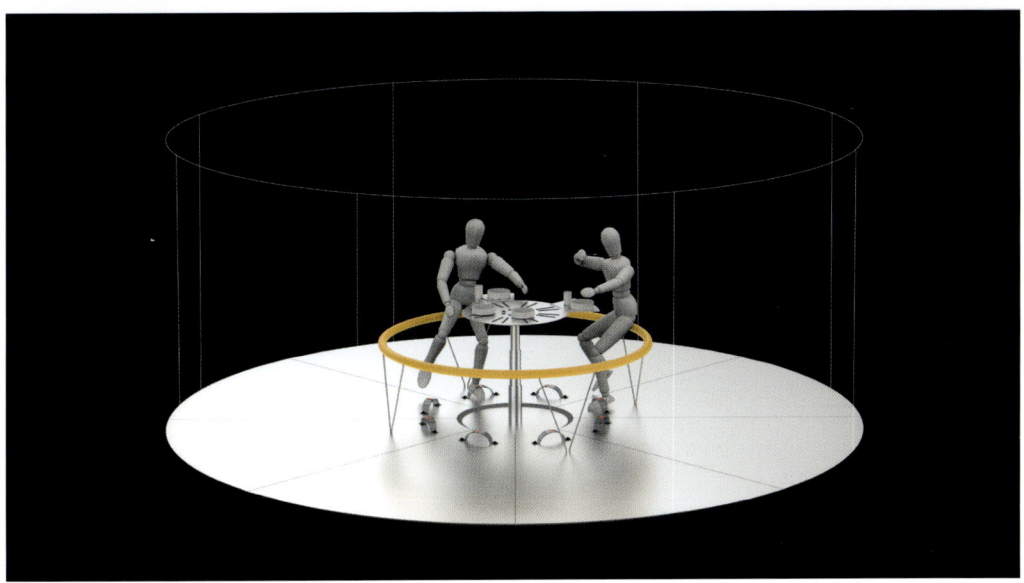

There have been numerous attempts to design tables for spacecraft. Currently, space stations have highly engineered tables or benches that serve both for meals and a variety of functional tasks. But the long, voyage to Mars requires much more careful design. With this in mind, the Design Museum commissioned a 'dining table' for the journey to Mars from the designer Konstantin Grcic. Grcic's first instinct was to do away with the over-engineered feel of these tables and design something simpler to use. Rather than seats, he provides a rail that astronauts can rest against, with their feet hooked into floor straps to stop them floating away. The table also sinks into the floor to free up space. Grcic's structure gives astronauts a sociable space that encourages the valuable ritual of communal meals.

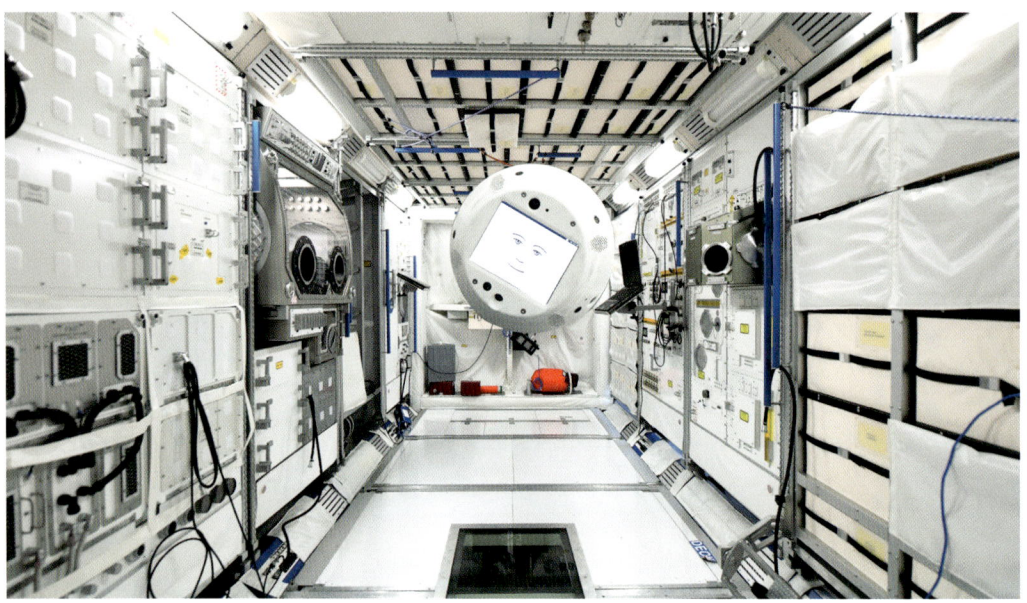

German Aerospace Center (DLR), composite image of CIMON (Computer Interactive Mobile Companion) on the ISS, 2018. A head-shaped AI robot that assists crew members on the ISS.

SURVIVE

MISSION CRITICAL: MARS MODERN
Fred Scharmen

A building is like a soap bubble. This bubble is perfect
and harmonious if the breath has been evenly distributed and
regulated from the inside. The exterior is the result
of an interior.

Le Corbusier, *Towards a New Architecture*[1]

Paul Rudolph, Crawford Manor
exterior, New Haven, CT, 1962-6.

Designers have often used other planets as sites for which
they could imagine new kinds of spaces. Uncoupled from
the heavy histories of Earth's architecture, new worlds
are places where new technologies and new spatial models
can be explored and experimented with. At the same time,
buildings in outer space have a very basic mission: they
must keep the humans inside alive. This tension between
knowns and unknowns in the off-world architectural
imagination is not without historical precedent.

     Almost a hundred years ago, Le Corbusier was thinking
about how new technology might change the way architecture
makes space on Earth. His principles for this new way of
working would influence the way buildings were conceived
in the twentieth century. The seemingly inevitable rationality
of this new Modernist architecture connected with the logic
of industrial production and construction, and appealed
to institutional and corporate interest in understanding and
rationalising large territories and populations. But the true
test of a Modernist building's success or failure was how
well it functioned. In Le Corbusier's grand system – famously
– a house is a machine, but also a street is a workshop and
a skyscraper is an instrument. Together, these components
comprised a vast mechanism that spread out across the
landscape of the Earth, claiming and globalising new space.
In the system of Modernism, form follows function, but the
functional imperative of individual buildings hides the larger function of
the system itself – Modernism conquers worlds.

     By the 1970s, this grand mechanism was showing signs of wear
and weakness. Some critics pointed out that the heroic rationality of
Modernism had failed to communicate anything to ordinary people.[2]
Others maintained that Modernist architecture's relationship with public
government services – especially public housing – had failed to function
in serving the needs of the vulnerable.[3] Still others noted that, even
though it had become the dominant mode in twentieth-century city-
making, Modernist architecture had failed in at least one crucial urban
task: it couldn't seem to shape coherent outdoor space in cities.[4] One way
or another, these failures in communication, service and space-making
were all symptoms of the same thing: having conquered the old world,
Modernism had been unable to build a new one.

     But Earth is not the only world. In proposed off-Earth architecture,
rationalist and functionalist imperatives are back – alongside heroic, world-
scale ambition – with several twists. There are new chances to build new

relationships with worlds here, and new opportunities to do a better job of working with existing ones.

Recent proposals for human habitats on Mars show a new architecture based on a renewed interest in old criteria. These projects argue for their own existence in terms of their engagement with functionality, performance and construction methods. The project proposals for the 3D-Printed Habitat Challenge, organised by NASA with several industry partners, might not look like the mid-twentieth-century design work we associate with architectural Modernism, but they rely on the same kinds of framing. MARSHA, the winning proposal by AI SpaceFactory, makes a strong argument rooted in an integration of form

and function. Their tower organises working, sleeping and recreational space vertically, suggesting that life itself takes precedence in the hierarchy of needs. The narrowness of the simple form also presents an optimally small footprint for the construction robot to work in, and performs efficiently as a pressure vessel to hold air. Proposals by HASSELL in partnership with Eckersley O'Callaghan and by Foster + Partners both take a series of capsules, differentiated by function, and unify them under a printed dome – like rocks snuggled under a blanket – to protect them from cosmic radiation. The Mars X-House V2 by SEArch+ and Apis Cor fuses structural, safety and programmatic factors into a complex whole, resembling a cooling tower with an egress stair wrapped around it.

AI SpaceFactory, MARSHA cross section and 3D views, render, 2018.

These functional factors are, of course, of utmost importance in this distant, hostile environment, where every square metre of habitable interior space is expensive to produce, and unprotected exposure to the exterior is fatal. But the form of AI SpaceFactory's MARSHA is recognisable as more than just an index of its function. Besides giving, the designers say, more usable interior space than a low dome would, it also has a striking silhouette – standing out like a vertical beacon in the Martian landscape, instead of hunkering down and hiding horizontally. This new Modernism

shows that it has learned a lesson from the failures of the old: it must explicitly communicate. This communication takes place within many contexts. The first crewed mission to Mars, and the first opportunities for humans to live there, would be a triumph of applied science and engineering. But before they become detailed plans, events like this are rehearsed in the human cultural and visual imagination. One of the unspoken tasks of the old Modernism was to translate and mitigate the disruptive power of technology with aesthetics. The sublime multi-level cities of Le Corbusier were akin to schemes by contemporaries like Frank R Paul, illustrator of early twentieth-century serials *Amazing Stories* and *Marvel Comics*, or Fritz

AI SpaceFactory, MARSHA in the Martian landscape, render, 2018.

Lang, director of the 1927 blockbuster science-fiction film *Metropolis*. Similarly, mid-twentieth-century science-fiction scenarios for missions to Mars, illustrated by architect and film-concept artist Chesley Bonestell, and published in magazines like *Life* and *Collier's*, were the inspiration for a whole generation of space and rocket scientists. So these new structures on Mars should communicate as new terms in the history of science and the history of architecture, but also engage with the history of speculative and science fiction, which is intricately bound up with both. These new proposals for Mars, created for the 3D-Printed Habitat Challenge, succeed to the extent that they take their role as cultural communicators seriously, and address for the future science-fictional questions about space.

One key question about this kind of space: who is it made for? Modernism sought to create a universal system, designed around an imaginary universal human subject. These aspirations are maybe best figured by Le Corbusier's famous Modulor Man, the basis for a proportional geometry he used to determine everything from countertop levels to building heights. He decided to base these proportions on a six-foot-tall human, but not just any human: 'Have you never noticed that in English detective novels,' he wrote in his book *The Modulor*, 'the good-looking men, such as the policemen, are always six feet tall?'[5] This is a universal system of measurement begun with the height of an imaginary male European authority figure. In space – in architectural space, and in space architecture – everything that is designed will end up inviting certain people to use and occupy it, and (perhaps unintentionally) disinvite others. Architects relearned this lesson when it was discovered, in 2015, that building-environment standards for air temperature in workplaces were based on science from the 1960s, which assumed that most people in offices would be men, with high metabolism, wearing business suits.[6]

AI SpaceFactory, MARSHA interior, render, 2018.

On Earth, we take the air for granted; not so on Mars. Its composition, temperature, pressure and humidity will have to be precisely specified there, along with key dimensions and all other human factors. Opening a window won't be an option, and neither will sending for a new spacesuit. If the built spaces on Mars proliferate, and start to diverge from one another in their specifications, then, for better or worse, so might the humans who feel comfortable within them.

In our human worlds, public space is where different kinds of people can, in theory, come together to share things: experiences, backgrounds, stories, existence itself. Public space is where *difference* can be appreciated and valued, and the built environment builds culture and community out of it. The perceived failures of Modernist planning to create legible public space gain a new resonance in off-Earth environments, where there is effectively no occupiable exterior. The low air pressure, lack of oxygen and freezing temperatures on Mars would kill an unprotected human in less than a minute. Here, to extend Le Corbusier's notion of the building as a soap bubble, the exterior is not just the result of an interior: it is simply *more* interior. In the most recent generation of proposals for buildings on Mars, excursion suits are attached directly to the wall. To take a walk on Mars, you would first step into your suit from inside the habitat, and then seal it off and walk away. Instead of using the classic airlock that cycles back and forth between exterior and interior, this is as if the soap bubble were cleaving off tiny copies of itself into the world at large, and reabsorbing them on their return.

In September 2017, Elon Musk presented a long-term plan for building on Mars to the International Astronautical Congress meeting in Australia.[7] The last phase of that plan was an image of a city composed of an assortment of modules of different sizes and shapes, sprawling across a rough grid. On Mars, public space must be designed as interior space or not at all. Earlier versions of cities on Mars, especially those illustrated by Bonestell, were organised around large, central domes, with clear density and hierarchy, and clearly legible public space under one all-encompassing envelope. Musk's city has a different kind of form, and it implies a different kind of social existence: more individualistic and even solipsistic. If we are all isolated in our own bubbles, we are not sharing space.

Modernism was bad at making outside space because it didn't recognise externalities. Modernist architects resisted recognising that their work was more than just a pure translation of function to form, that it was communicating in different cultural, social and political registers, whether they intended it to or not. They largely didn't acknowledge difference, in human culture, gender, ability or physiognomy. And they didn't make coherent space for that difference to be recognised and celebrated. Despite their functionalist foundations, some of the proposals for Mars I have discussed express a willingness and ability to deal with concerns that were disregarded by the Modernists. Instead of a globalised logic of mass production, the 3D-Printed Habitat Challenge shows that architects could build custom shapes and environments, using material that is already on site. Instead of producing many small bubbles with an inefficient use of surface area, larger enclosures might be more cost-effective, and could encourage shared cultures.

Shared space, in hostile environments, also hints that basic resources might also be shared – air, water, energy. Why not food? Healthcare? Education? Instead of isolating the discipline of architecture as something autonomous of social, cultural or even fictional realms, we might embrace and celebrate the interconnections between architecture, politics, space science and science fiction. In her book *Placing Outer Space: An Earthly Ethnography of Other Worlds* (2016), anthropologist Lisa Messeri writes about the concept of a 'planetary imagination'.[8] This is a category that allows for broad thinking about the relationships between worlds and people. On Earth, it is architecture that mediates those relationships and expresses that imagination. Classical twentieth-century Modernism is one kind of planetary imagination, one that sees worlds as things that can be rationalised by an approach to architecture that flattens difference and imposes its own logic on new territory. Mars is not a new world – it's an old one. Four billion years ago it may have been not so different from Earth. Mars has been many planets in the past, and Earth may end up being many other kinds of planet in the future. We will need new kinds of building for all of these worlds, on Earth as much as elsewhere. Habitat projects for Mars are hints that other kinds of architectural planetary imaginations might be possible.

1   Le Corbusier, *Towards a New Architecture* (1931; repr., New York: Dover, 1986), 181.

2   Robert Venturi, Denise Scott Brown and Steven Izenour gently mocked the grand aesthetic effects of late Modernism as 'heroic and original', while suggesting that 'society' preferred buildings that spoke to them in ways that were more 'ugly and ordinary'. See Robert Venturi, Denise Scott Brown and Steven Izenour, *Learning From Las Vegas* (Cambridge: MIT Press, 1972), 128-9.

3   See Charles Jencks on the demolition of Minoru Yamasaki's Pruitt-Igoe public-housing project: 'Modern Architecture died in St Louis, Missouri, on July 15, 1972 at 3:32p.m. (or thereabouts), when the infamous Pruitt-Igoe scheme, or rather several of its slab blocks, were given the final coup de grace by dynamite', see Charles Jencks, *The Language of Post-Modern Architecture* (revised enlarged edn; New York: Rizzoli, 1977), 9.

4   See especially Colin Rowe and Fred Koetter, *Collage City* (Cambridge, MA: MIT Press, 1979).

5   Le Corbusier, *The Modulor: A Harmonious Measure to the Human Scale, Universally Applicable to Architecture and Mechanics* (first published in two vols, 1954, 1958; Basel and Boston, MA: Birkhäuser, 2004), 56.

6   Joost van Hoof, 'Building Emissions: Female Thermal Demand', *Nature Climate Change*, 5 (2015).

7   Elon Musk, 'Making Life Multiplanetary' (2017), www.youtube.com/watch?v= tdUX3ypDVwI [Accessed 24 May 2019].

8   Lisa Messeri, *Placing Outer Space: An Earthly Ethnography of Other Worlds* (Durham, NC: Duke University Press, 2016), 12.

The Mars Society, Flashline Mars Arctic Research Station (FMARS) in the Canadian Arctic, 2017.

The Mars Society, Jonathan Clarke and Anastasiya Stepanova during the Mars analog mission, Mars160, 2017.

The Mars Society, Crew 1-6 FMARS mission patch, 2001.

The Mars Society is a non-profit organisation dedicated to promoting human exploration, and eventual settlement, on Mars. Set up by Robert Zubrin in 1998, the organisation runs the Mars Analog Research Station Program, in which prototype Mars habitats are deployed and tested in Mars-like environments on Earth. Two stations are currently in operation – one in the Canadian Arctic and the other in the Utah desert.

The Mars Society, Mars Desert Research Station (MDRS) habitat, Utah, 2001 (photographed 2017).

The Mars Society, Crew-189 at MDRS, 2018.

University of Hawaii at Mānoa/NASA, Oleg Abramov on an EVA outside the Hawaii Space
Exploration Analog and Simulation (HI-SEAS) habitat, 2013.

University of Hawaii at Mānoa/NASA, interior of the HI-SEAS habitat, 2013.

Foster + Partners, Mars Habitat, axonometric drawing, 2015.

Foster + Partners, the deployment and inflation of Mars Habitat modules, render, 2015.

The scheme for Mars housing devised by Foster + Partners envisages using robot builders sent out from Earth in advance of the astronauts. The design of these robots poses a significant challenge as they will need to be semi-autonomous and intelligent - responding to their environment independently, as much about the Mars surface and subsoil is unknown and can therefore not be programmed for.

Foster + Partners, Mars Habitat, Operational Mars Base, render, 2015.

Foster + Partners, interior laboratory module of the Mars Habitat, render, 2015.

NASA and University of Arizona, Dunes in Aonia Terra, Mars, 2012. Images taken by the High
Resolution Imaging Science Experiment (HiRISE) camera on board the Mars Reconnaissance
Rover. HiRISE takes extremely high-resolution images of the surface of Mars in order to
aid the identification of optimum sites for landing and future settlements.

Mars City Design and Alpha Team (Vera Mulyani, Stewart Davis, Paola Vezzulli, Lara Hoad,
Aleksandar Bursac and Ivan Djikanovic, Ilaria Campi, Jasmine Park), Alpha 2.0 (Parametric
Barchan Dunes Pocket Studies), 2018-9. All Mars dwellings will need a substantial amount
of structure to shield their occupants from radiation. Like other schemes, Alpha 2.0 proposes
fusing or printing the regolith sand, or topsoil, to make the initial structures of the
habitats, but these are intended as seeds that will encourage the natural processes by
which sand dunes form on the planet. They will be located in an active dune field, Nili
Patera, near the Martian equator. Over time, the windblown topsoil will eventually envelop
and cover the habitats with a thick protective layer.

SEArch+ (Space Exploration Architecture) / Clouds AO (Clouds Architecture Office), Mars
Ice House, exterior, 2015. This design won NASA's Centennial 3D-Printed Challenge in 2015,
a competition to build a 3D-printed habitat for deep space exploration. It proposes using
ice water on Mars to 3D-print a translucent habitat for four explorers. In 2019, SEArch+
designed another award-winning proposal: the Mars X-House (see pp. 144-7).

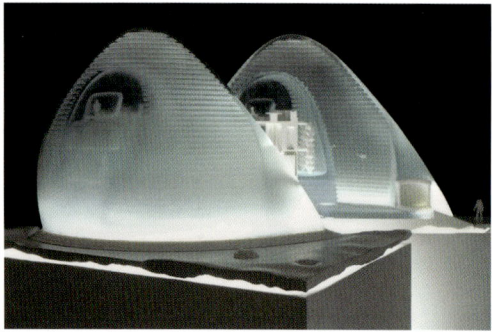

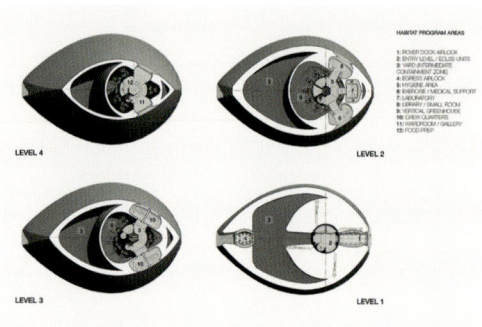

SEArch+ (Space Exploration Architecture) /
Clouds AO (Clouds Architecture Office),
Mars Ice House, scale model, 2015.

SEArch+ (Space Exploration Architecture) /
Clouds AO (Clouds Architecture Office),
Mars Ice House, plans, 2015.

SEArch+ (Space Exploration Architecture) / Clouds AO (Clouds Architecture Office),
Mars Ice House, section perspective, 2015.

SEArch+ (Space Exploration Architecture) / Clouds AO (Clouds Architecture Office),
Mars Ice House, exterior, 2015.

SEArch+ (Space Exploration Architecture) / Clouds AO (Clouds Architecture Office),
Mars Ice House, front yard (left) and wardroom (right), 2015.

SEArch+ (Space Exploration Architecture), Mars X-House, exterior, 2019. This design was the first-prize winner of NASA's Phase 3 3D-Printed Habitat Challenge in Virtual Design.

SEArch+ (Space Exploration Architecture), Mars X-House, exterior, 2019.

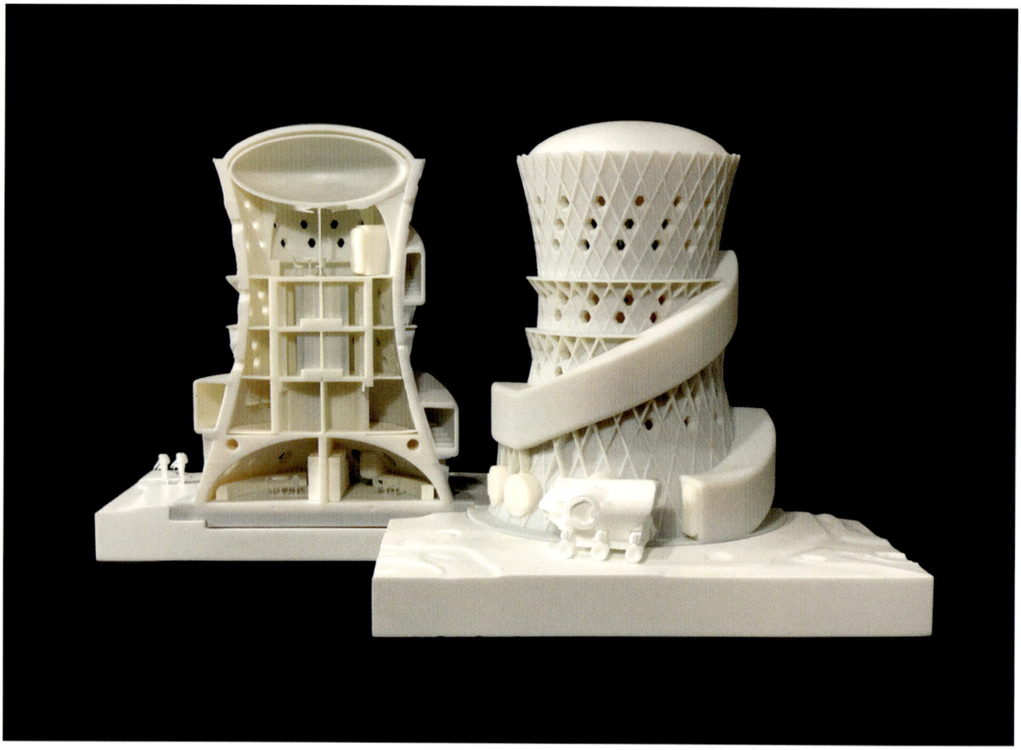

SEArch+ (Space Exploration Architecture), Mars X-House, cross section model, 2019.

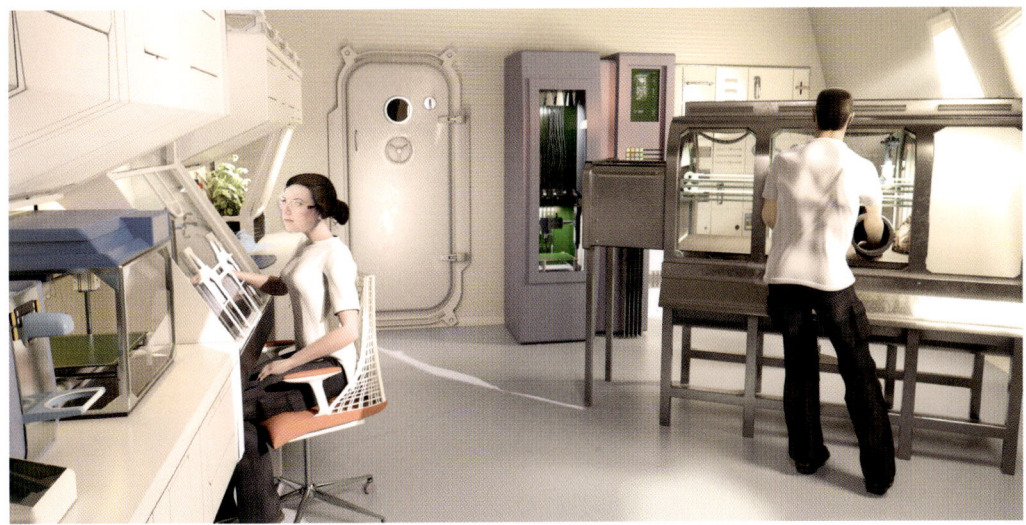

SEArch+ (Space Exploration Architecture), Mars X-House, laboratory, 2019.

SEArch+ (Space Exploration Architecture), Mars X-House, wardroom, 2019.

The Mars X-House by SEArch+ is designed to minimise the cosmic and solar radiation that its inhabitants are exposed to. The parts of the dwelling that will be least used, such as the stairway, are placed outside, and the upper floor is used as a water tank, with the water blocking radiation while still allowing light to enter the habitat.

HASSELL, 3D-printed habitat, render, 2019. An inflatable habitat protected from the Martian
environment by a shield formed from regolith.

HASSELL, 3D-printed habitat, render, 2019. Beneath its regolith shield, the habitat is made up of inflatable pods that are connected by smaller ones which provide life-support systems for its inhabitants.

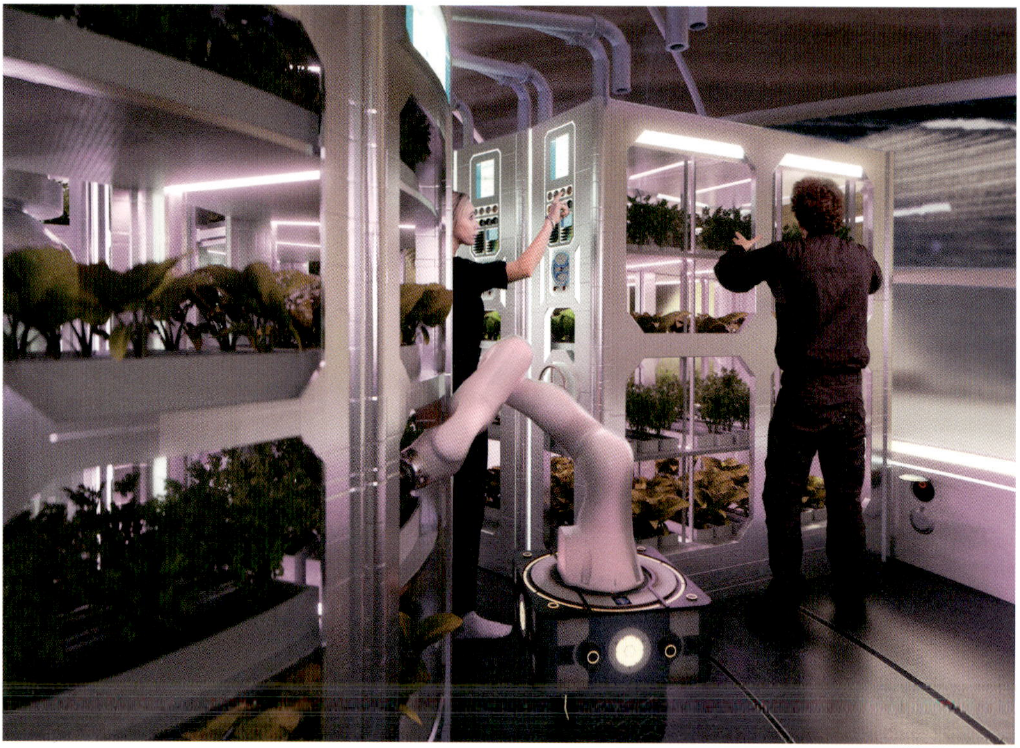

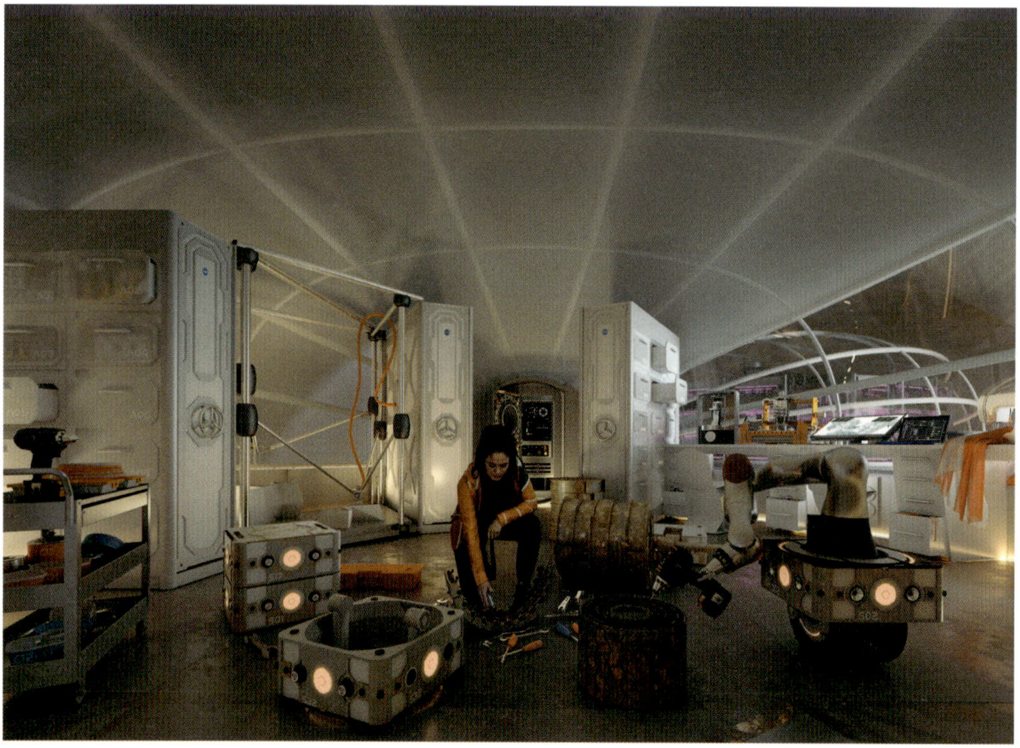

HASSELL, 3D-printed habitat interior views, renders, 2019. Clockwise from top left:
greenhouse; living spaces; gym; research laboratory with movable rack system. Garments
designed by RÆBURN.

RÆBURN, Spring/Summer 2020 NEW HORIZONS Collection, 2019.

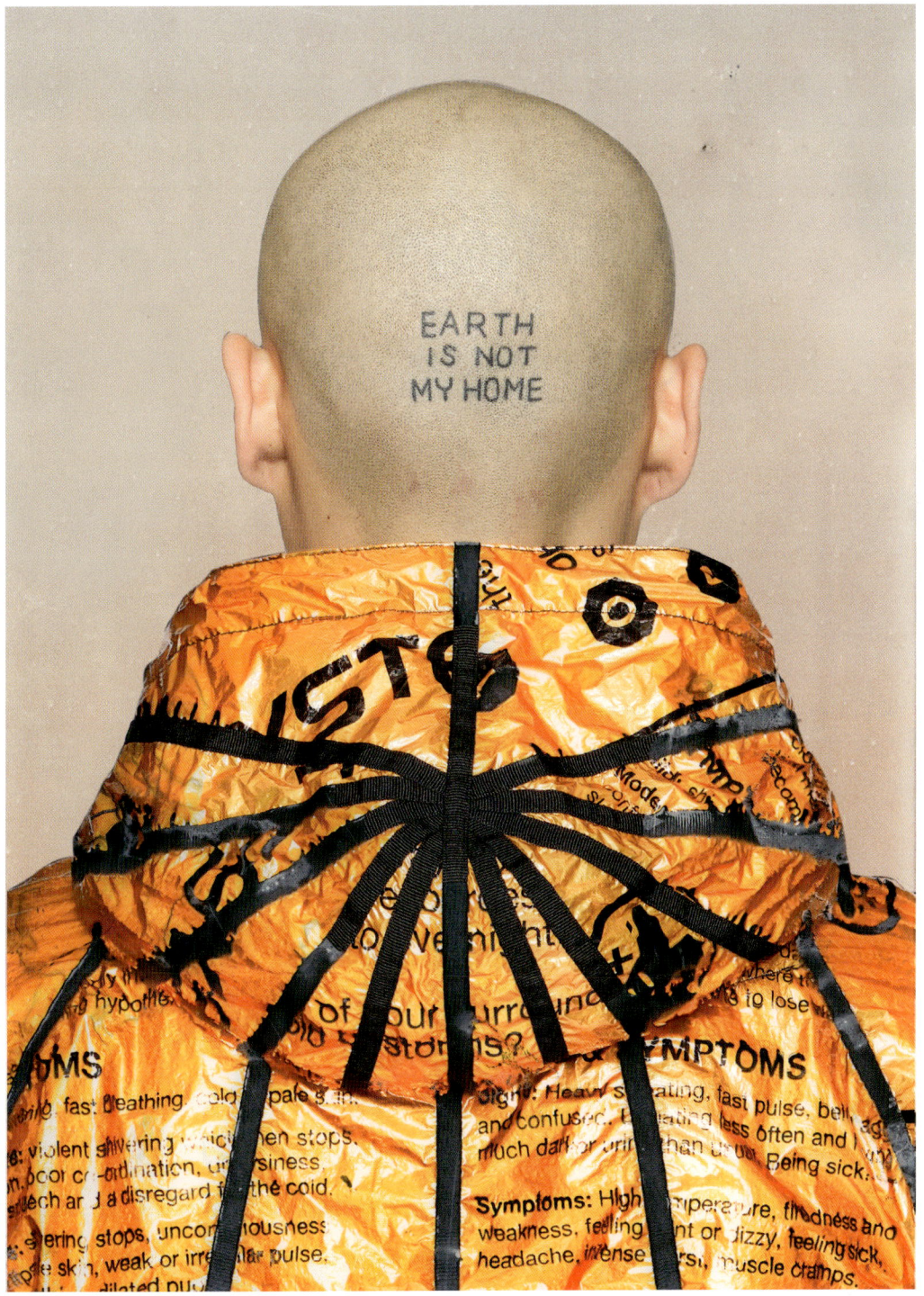

For its Spring/Summer 2020 NEW HORIZONS Collection, RÆBURN collaborated with architecture firm HASSELL to create garments that explore the needs and restrictions of future astronauts landing on Mars. Made from recycled parachute material and solar blankets, the collection explores the 'make do and mend' attitude that will be needed to live on the planet, where all materials will have to be intelligently reused and recycled.

Katie Paterson, *Timepieces (Solar System)*, 2014. A series of nine clocks that tell the time
on the planets in our solar system and Earth's Moon. The durations of day and night range
from planet to planet, from the shortest on Jupiter to the longest on Mercury. Each clock
is calibrated to tell the time in relation to the other planets and to Earth. Left to right
from top: Mars, Mercury, Earth, the Moon, Jupiter, Saturn, Uranus, Neptune, Venus.

CONTAMINATING THE RED PLANET
Lydia Kallipoliti

'We live in a dirty world', claimed NASA's Office of Technology Utilization in 1969.[1] The presence of microscopic dust particles could not be tolerated, as they could clog up machinery, choke filtration systems and ultimately compromise manned missions. Since the dawn of the Space Age, the sterilisation of space probes has been a necessity, not only to prevent equipment malfunctioning, but also to prevent terrestrial life from interfering with native life on other planets. Scientific investigations to detect life in the universe could be jeopardised if we are not vigilant in protecting foreign planets from contamination by any Earth organisms that might hitchhike on a spacecraft, survive the trip and grow to multiply on other worlds.[2]

The problem of planetary contamination was first raised at the International Astronautical Federation VIIth Congress in Rome in 1956, which led to the formation of the committee on Contamination by Extraterrestrial Exploration (CETEX).[3] As explained by Lawrence Hall, an interplanetary quarantine officer at NASA in the 1960s, 'if we change the biology of a planet in some way, or introduce new life where none existed before and it proliferates, we could change many of the characteristics of the planet'.[4]

Because of this, plans for a manned Mars Excursion Module (MEM) in the 1960s involved astronauts, on landing on Mars, ejecting shields that protected MEM windows from local hazards, including 'unfriendly life forms'.[5] With sterilisation protocols in place, Mars has been understood mostly as a biological preserve, or a laboratory in its purest form, with rigid protocols to remove all contaminants. The planet itself has been seen as an untouched, pure territory, an experimental blank terrain, with nothing to interfere with or interrupt the experiment. No forward or backward contamination of living tissues has been allowed or planned. Once – and if – Mars is inhabited, however, it will be contaminated by other microorganisms. It will, by default, no longer be a preserve. But, most significantly, any type of colonisation is also an infection – a blending of substances, species and materials from either biogenic or abiogenic sources.

Philco for NASA, artist's illustration of the proposed Mars Excursion Module (MEM), 1964.

The colonisation of Mars will inevitably demand the separation and protection of humans inside a closed life-support system: a self-sustaining physical environment separated from its surroundings by a boundary that does not allow the transfer through it of matter or energy. As partial reconstructions of the world in time and in space, closed systems identify and secure the cycling of materials necessary for the sustenance of life. Contemporary discussions about global warming, recycling and sustainability have emerged from the study and analysis of closed systems. First, they enable scarce resources – water and oxygen – to be reused and recycled by being extracted, filtered and recirculated; most importantly, though, closed systems convert waste into new viable commodities. Nevertheless, engineered closed-loop systems, which are mostly portrayed in simulations as robust circular systems where waste equals food in an endless series of cycles and subcycles, are fragile and vulnerable because they depend on subtle biological interactions and the digestion of resources. Contaminants generated by both human occupants and the system's own component parts and materials can compromise and impair their function. Indeed, the instability and uncertainty of artificial closed systems demonstrates the density of life and the complexity of ecosystems on Earth.

In the late 1960s, as the image of the whole Earth and the effects of the space programme began to have an impact on the cultural imagination, it was suggested that the hardware from spacecraft could be directly employed by building industries on Earth, yielding ecological benefits and transforming buildings into life-support systems that recirculated all their resources. Certain life-support systems that were experimented on in outer space were not successful in zero-gravity conditions, due to the difficulty of controlling the flow of bacteria and microzoa that were necessary for the anaerobic digestion of waste products. Problems that originated with bacteria flocks under gravity-free conditions in space, however, did not present the same difficulties when the same processes were tried out in residential settings back on Earth.[6]

Developed in the years after the Space Race, the largest and most famous closed ecological system ever built on Earth was Biosphere 2 in Oracle, Arizona. Its purpose was to test the viability of a biologically regenerative artificial environment that might be able to support human habitation in space. The idea was that it could provide a prototype spaceship – or fallout shelter – to preserve a habitable place in readiness for the inevitable destruction of Biosphere 1 (the Earth). Biosphere 2 supported two missions, with the first team of researchers entering the facility on 26 September 1991, for two years. The team produced their own food and air supply within a heavily sequestered and maintained series of ecosystems. During the entire experiment, all of the crew's waste, including that from their domesticated animals, was recycled using natural, low-tech filtration methods. Despite the intense preparations for the manned mission, Biosphere 2 failed spectacularly and publicly.

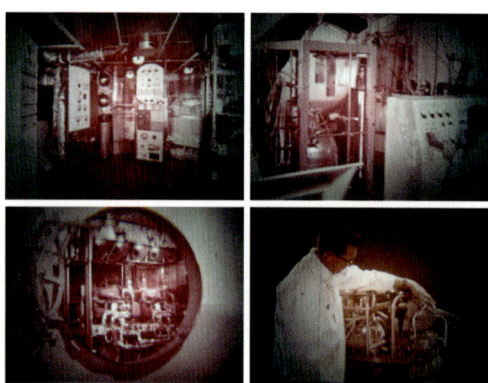

NASA, *The Case for Regeneration* film stills, 1960.

The first mission was notorious for technical and operational failures, especially for not managing to create enough oxygen. After six months, oxygen absorption by raw concrete that was still curing had depleted oxygen levels in the facility by six per cent. That deficiency, combined with the need to dump tonnes of sodium bicarbonate in the facility to offset the pH imbalance in its ocean biome, led the scientific community and popular media to deride the integrity of the project. Along with poor interior air quality, unanticipated species like cockroaches and 'crazy ants' thrived, and the biospherians suffered from hunger throughout the two years of their residency.

The unpredictable developments in Biosphere 2 are representative of closed systems, which operate in many instances like 'black boxes', producing new substances that were unaccounted for in the system's design. In fact, the minute a human body enters the closed-loop system of any given space habitat, the flawless operation of its feedback loops are disrupted. A good illustration of this fragility is provided by NASA Langley's simulator, known as the Living Pod. In 1960, four subjects were enclosed in it for four months, entirely sequestered from the outside world. Throughout the course of the experiment, the subjects experienced nausea and headaches, and eventually contaminated the system with their own waste. Shed hair, fingernails and skin infiltrated the collection systems; eventually, the subjects had to be removed from the cabin earlier than expected. In addition to carbon dioxide, contaminants in the closed ecosystem included minute waste particles like dust, hair, skin debris, tobacco particles, odours and toxic substances from cooking, and other formed organic compounds with unpleasant odours like spores, viruses and saprophytic bacteria that decompose organic matter. The probable extent of the contamination was not recognised until this particular living experiment took place, primarily because there had been such a high leakage rate in earlier spacecraft. Once the occupancy took place on the ground, in the simulator, more than 400 contaminants were identified. Even the simulator's materials – paints, coatings, plastics and insulation materials – gave off vapours, among them benzene, formaldehyde and acetone. The contaminants produced by organic processes, however, were even more harmful. Metabolic processes produced methane, carbon monoxide and ammonia.[7]

Therefore, the image of astronauts as heroic explorers who overcome their physiological boundaries and conquer uninhabitable lands obscures problematic material realities. The cybernetic feedback diagram – familiar as a flow chart since the 1960s – has failed to predict the fact that the pod, as a living system, itself produces new matter. The feedback diagram, with all its multipart recursion loops, cannot account for the partial local material bodies that are formed at a smaller scale. Eventually, such newly formed bodies destabilise the system, and disrupt the balance of closed habitats.

The Martian, directed by Ridley Scott, film still showing Matt Damon as Mark Watney on the surface of Mars, 2015.

Geopolitically, the Martian astronaut, hovering in a mask and gear over a territory unfriendly to the physiology of humans, propagates the iconography of the heroic settler.[8] When looking at Mark Watney (played by Matt Damon in Ridley Scott's 2015 film *The Martian*), the fictional character left behind on Mars,[9] against the vast barren landscape of the Red Planet, there are clear reflections of Caspar David Friedrich's *The Wanderer above the Sea of Fog,* a painting that defined the Romantic period and the iconography of the sublime. This visual narrative of the explorer as the inevitable victor in a free terrain pervades our earthly history of colonisation, in which power differentials and conflict have been established by the demarcation of dominion.

Colonising Mars begs the question of where the body is located in this process: it is no longer a figure seen from afar in a desolate landscape, but a corporeal physical entity that is no longer outside the biological preserve of the planet. The human body would be the source of the planet's contamination: the cause of dirt. As a machine of ingestion and excretion – but also as the primal operator of labour and daily tasks – the astronaut is both an experimental subject and the monitoring agent of the experiment itself. The Martian explorer, therefore, far from a captor, is a scientist and a guinea pig, keen to insert their body into the experiment and to offer it as a test bed for further trials. In this light, bodies are no more than expressions of change, carefully mapped and monitored. Within the premises of closed-loop habitats that will ensure survival, bodies are reorganised, unpacked and distributed in space to yield other resources. This type of dissemination requires a radical shift in our relationship with our bodily excrement – in our relationship with our own shit, both literally and figuratively.

The handling of our own excrement within a closed life-support system for Mars forces us to look at questions of space colonisation viscerally, through the raw ecology of our bodies. Recycling is, in this context, not merely a mathematical equation of feedback loops and flow charts, but also a basic bodily reality that affects the water and air available for survival on this so-called 'new frontier'.

In the late 1970s, Captain Robert Freitag, deputy director of the Manned Space Flight Center at NASA, declared that much was still unknown about the interactions that take place in a closed ecosystem. He proposed algorithms should be developed to define the basic supporting relationships between humans, animals, plants and microorganisms, in order to define the conditions under which ecological closure might exist. This would prove to be the single most demanding technology to be developed in the twentieth century.[10] After years of experimentation with ecological closure, biologists at the time came to similar conclusions: despite the rigour of mathematical formulas, contained artificial ecosystems were unpredictable in their evolution.[11] If even minor ruptures occurred in any part of the system, closed worlds had no 'healing mechanism'.

Closed-loop habitats are digestive machines, and they are sometimes disobedient: this is what self-reliance and autonomy mean in many respects. Even though life-support systems are mostly framed in simulations as being robust circular systems, where waste equals food in an endless series of cycles and sub-cycles, the reality is that self-sufficiency is compulsive in its desire ceaselessly to generate new life from all wasteful cycles of production. These different types of new life, formed from our own bodily debris, will become our companions in the highly engineered, hermetically sealed, closed-loop habitations in which we will make our lives on Mars.

1   James W Useller, *Clean Room Technology* (Cleveland, OH: Lewis Research Center; Washington, DC. Office of Technology Utilization, National Aeronautics and Space Administration, 1969), 5.

2   Letter from Joshua Lederberg, University of Wisconsin, to Detlev Bronk, President, National Academy of Sciences, 24 December 1957, with enclosed memorandum entitled 'Lunar Biology?', National Academy of Sciences, Records Office, Washington, DC.

3   CETEX was founded in 1959, and later superseded by the Committee on Space Research (COSPAR).

4   Lawrence B Hall, 'Sterilizing Space Probes', *International Science and Technology* (April 1966), 50-61.

5   Franklin Dixon, 'Summary Presentation: Study of a Manned Mars Excursion Module', *Proceeding of the Symposium on Manned Planetary Missions: 1963/1964 Status* (Huntsville, AL: NASA TM X-53140, 1964), 443-523.

6   Dan C Popma and Vernon G Collins, 'Space Vehicle Water Reclamation Systems, A Status Report', in the *Chemical Engineering Progress Symposium Series*, 62/63, 1966, 5.

7   Howard W Mattson, 'Keeping Astronauts Alive', *International Journal of Science and Technology* (June 1966), 28-37.

8   Peter Redfield, 'The Half-Life of Empire in Outer Space', *Social Studies of Science* 32/5-6 (December 2002), 791-825. Available for download from doi:10.1177/030631270203200508.

9   See the film *The Martian* (2015), directed by Ridley Scott. USA: 20th Century Fox.

10 See Captain Robert F Freitag, 'Summary of Problems of Greatest Urgency', Princeton University Conference on Space Manufacturing Utilities, 1977. M1045, Series 1, Box 1, Whole Earth Access, *CoEvolution Quarterly* Records, Stewart Brand editorial files, Correspondence. Department of Special Collections, Stanford University.

11 See John Todd's response to Gerard O'Neill's 'Space Colonies', in Stewart Brand (ed.), *Space Colonies: A CoEvolution Book* (San Francisco: Waller Press, 1977), 48-9. For earlier reference, see Howard Odum's discussion in the *American Biology Teacher* 25 (1963), 423-43.

Lydia Kallipoliti with Jestin George, Beau Avedissian, Ka Hou Cheang, Dorsa Fahandezh, Jialu Huang, Mariam Mesiha and Isabella Wells, 'Life on Mars; From Feces to Food', 2019. An extra-oxygenated virtual reality graphene dome to project fantasies and memories for the crew. This project was supported by the University of Technology Sydney (UTS).

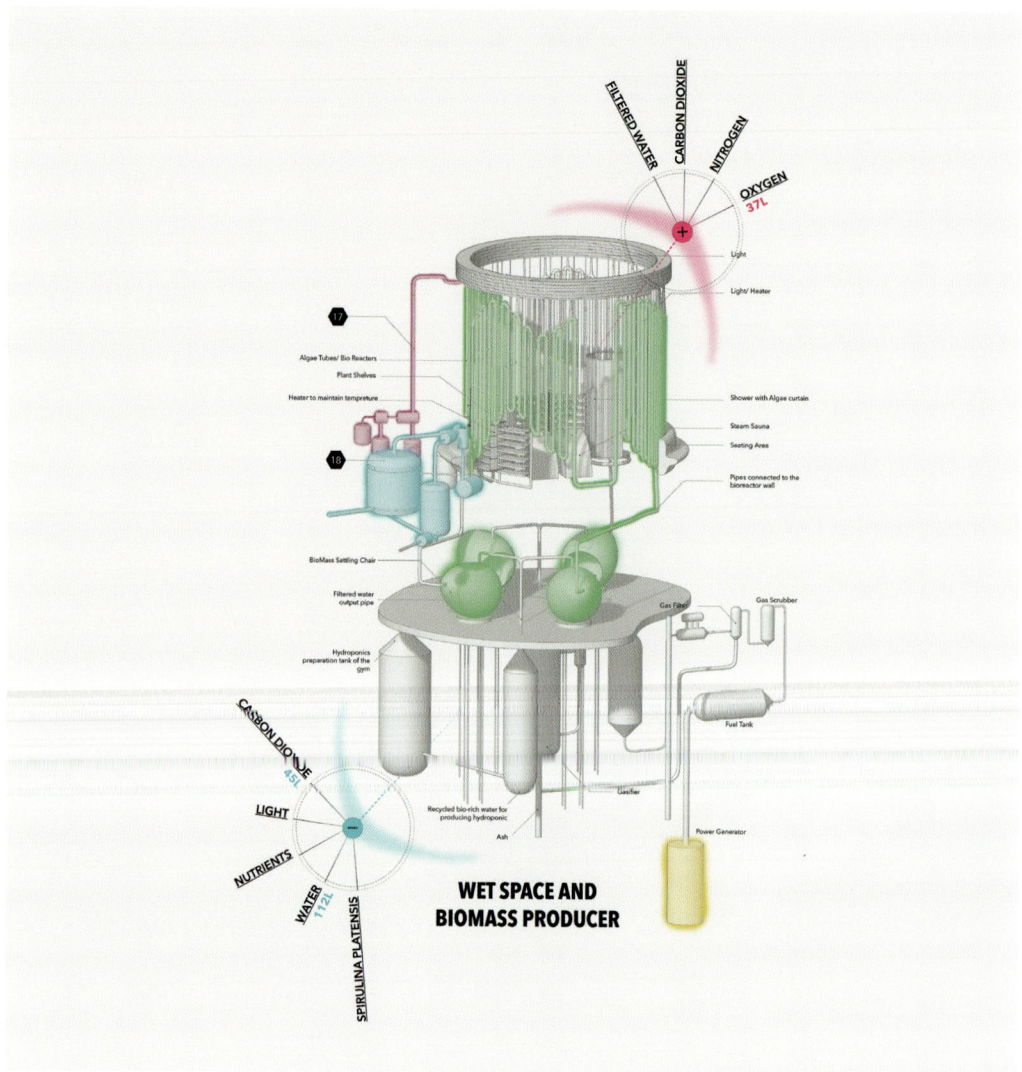

Lydia Kallipoliti with Jestin George, Beau Avedissian, Ka Hou Cheang, Dorsa Fahandezh, Jialu Huang, Mariam Mesiha and Isabella Wells, 'Life on Mars; From Feces to Food', 2019. Diagram of the 'wet space' in a hypothetical closed-loop system for Mars showing bioreactors with shower and sauna connected to the biomass-digestion lounge. This project was supported by the University of Technology Sydney (UTS).

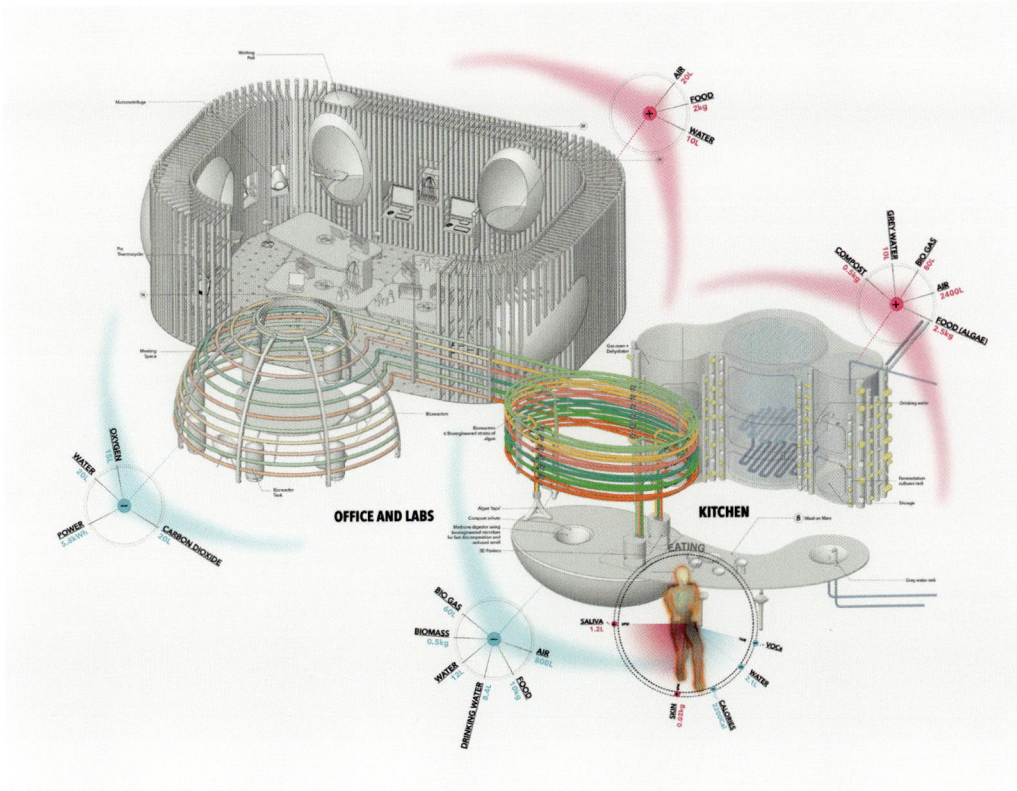

'Life on Mars; From Feces to Food', 2019. Laboratory kitchen where food is coded in a computer interface and eventually 3D printed in the kitchen through DNA synthesizers and bioreactors connected to the 3D printer.

Architect Lydia Kallipoliti has been studying 'closed worlds' like space capsules and submarines. In 'Life on Mars; From Feces to Food', she examines the complex closed-loop system that would be required to sustain life on Mars. Closed systems ensure that essential resources – particularly water and oxygen – are recycled. Equally importantly, they need to convert waste back into food. Engineering schemes often show these closed-loop environments made up of 'black box' accessories tacked onto the living quarters and linked in a flow chart where food converts to waste and back in an endless cycle. But these closed environments are fragile because they depend on subtle biological interactions. 'Life on Mars' is an attempt to design the engineering flow chart without resorting to black boxes and arrows. Instead, the machines that convert waste into viable resources are vital, lived-in elements of the habitat. Kallipoliti proposes new domestic pairings linking, for example, a toilet and a garden, a bioreactor and a bathroom, or a kitchen and a laboratory. 'Life on Mars' suggests a Martian colony that can almost be viewed as a life form in itself – a complex integrated system where humans and their innate physiology of ingestion and excretion become biological parts of the whole system that they inhabit.

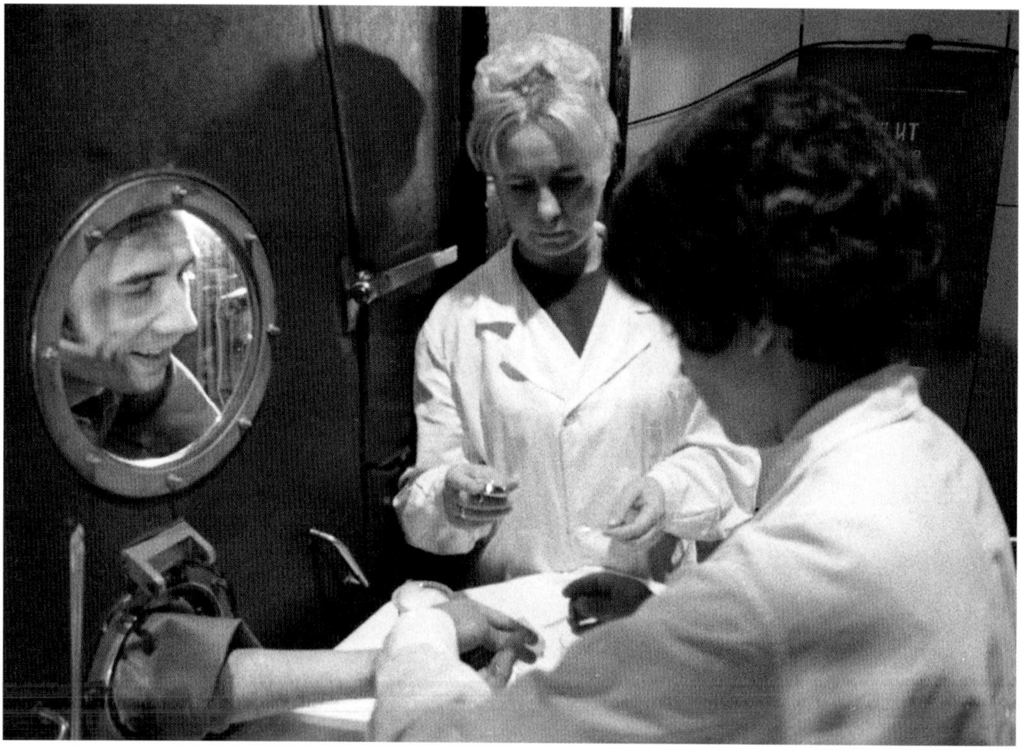

Russian Academy of Sciences, a BIOS-3 bionaut has his blood drawn for testing, 1973.

Russian Academy of Sciences, BIOS-3 bionauts Nikolai Petrov and Vladislav Terskikh harvest wheat ready for grinding in the facility's mill, 1973.

Russian Academy of Sciences, Bionaut Nikolai Petrov uses the negotiation console to communicate with the outside world, 1973.

BIOS-3 was a closed ecological life-support system run by the Soviet space programme from 1965 to 1972. Its aim was to develop a bio-regenerative life-support system for cosmonauts on the moon or Mars. The layout included four rooms: the crew's living quarters, an algae cultivator and two greenhouses. It was the first experimental ecosystem in which participants controlled the system from the inside, with three crew members living and working in the complex for six months.

Cutaway drawing of the human habitat at Biosphere 2, c. 1990.

University of Arizona, exterior of Biosphere 2, Arizona, 2008.

University of Arizona, Biosphere 2 ocean, Arizona, 2011.

Constructed between 1987 and 1991, Biosphere 2 is to date the largest and most notable closed-loop ecological system ever built. Spanning an area of 1.27 hectares and costing an estimated $200 million to build and maintain, the complex consists of five ecosystem regions – a desert, marsh, ocean, rainforest and savannah, all under glass domes. While the aim of the project was to prove the viability of closed ecological systems, its first two missions failed publicly and dramatically. The researchers living inside the complex had lost weight and suffered from lack of oxygen.

University of Arizona, Biosphere 2 rainforest, Arizona, 2011.

ESA, scientists working at MELiSSA, 1999.

Initiated in 1989, the Micro-Ecological Life
Support System Alternative (MELiSSA) is
an ongoing experiment by the European Space
Agency (ESA) looking to design a closed-loop
system for long-term manned space missions.

The project is run by a consortium of
thirty organisations, each investigating
a different aspect of food, oxygen and
water production.

NASA, the International Space Station's vegetable-production system 'VEGGIE' on display at Kennedy Space Center, 2014. An experiment studying the development of lettuce seedlings in a microgravity environment.

Limerick Institute of Technology and EDEN-ISS, plants growing in the EDEN-ISS Future Exploration Greenhouse (FEG), 2017.

China National Space Administration (CNSA), the Chang'e 4 lunar lander on the far side of the Moon photographed by the Yutu-2 Moon rover, 2019.

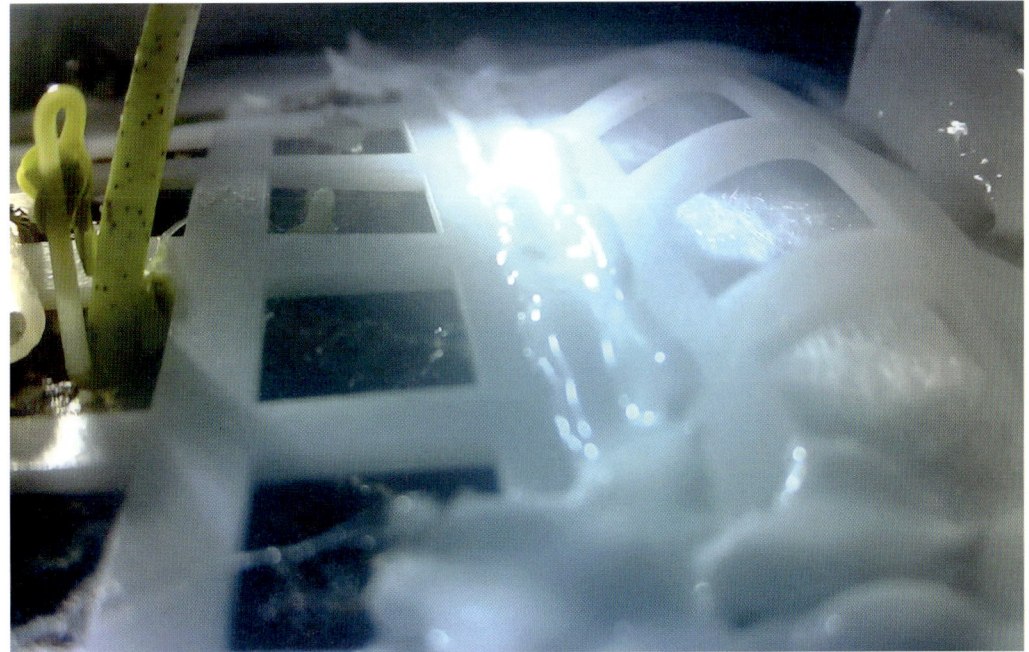

Center of Space Exploration, Ministry of Education (COSE), cotton seeds sprouting on the Moon inside a sealed and heated atmospheric module on board the Chang'e 4 lunar lander, 2019.

GrowStack, vertical farm, built for the Stockbridge Technology Centre, UK, 2018.

SPACE10, The Farm, an indoor hydroponic farm, 2016.

SPACE10, LOKAL, a hydroponic farming system at the London Design Festival, 2017.

AstroPlant, render showing LED light that encourages plant growth, 2019. AstroPlant is a prototype plant lab that allows people to collect data on potential crops to grow in space.

THRIVE

AN INTERVIEW WITH
KIM STANLEY ROBINSON
Justin McGuirk

Chesley Bonestell, *Domed colony on Mars*, 1976.

More than two decades after it was written, Kim Stanley Robinson's Mars trilogy – *Red Mars*, *Green Mars* and *Blue Mars* – remains the essential work of science fiction dedicated to the planet.[1] It is also one of the most convincing portrayals of human life in an alien context. That it should be a crucial reference point for this particular book owes much to the fact that Robinson believes in design – not just as a methodology for building literary worlds, but also as a way of making speculative ideas concrete. In the interview that follows, he describes design as 'a kind of grammar for living in the material world'. It stands to reason that a different world requires a different grammar, and much of the work in this volume is reaching towards one.

The tension at the heart of the trilogy is one between competing ideals, between irreconcilable utopian visions. The scientist Ann Clayborne is a preservationist who wants to study Mars in its pure state, and for whom any colonisation is a tragic despoiling. Arkady Bogdanov (who shares his last name with the Soviet novelist who imagined a socialist utopia on Mars in the 1920s) sees Mars as the seedbed of a revolutionary society built on egalitarian terms. And then there are the environmental transformers: Hiroko Ai, who eventually turns Mars green, and 'Sax' Russell, who terraforms the atmosphere to make it breathable.

The trilogy was an early sign that Robinson was to become one of the foremost novelists of climate change – as he has pointed out, we are already terraforming Earth, by accident. In *Blue Mars*, the symptoms of climate catastrophe on Earth include a flooded London. Ultimately, Robinson's Martian tale, with its new climate and teeming cities, extends far beyond most of the speculative design contained in this book. As he points out in the interview, Mars – if humans get there at all – is most likely to remain a scientific outpost for the foreseeable future, much like Antarctica.

And yet there is no shortage of utopian possibilities in that fact. Life on that barren planet will be basic and difficult, but, as Bogdanov proclaims in *Red Mars*, the 'real work' of being human is staying alive and satisfying one's curiosity. The twin tests of survival and discovery lie at the heart of Mars' appeal. As Bogdanov puts it, 'a scientific research station is actually a little model of prehistoric utopia'.[2]

JM   The reason why this exhibition is happening at the Design Museum, as opposed to a science museum, is that it is impossible to contemplate the nature of the human experience on Mars without design. Here we're not just talking about the shape of things, but the shape of experiences. Your *Mars* trilogy represents the most extensive piece of Mars world-building in literature. You had to visualise that world, from vehicles and habitats to, later, entire cities. How did you go about that? And how concerned were you with scientific plausibility?

KSR   I agree: design is crucial to the human project, and the more we understand that, the better we'll get at doing design – at designing our lives.

When writing my *Mars* trilogy, I was very concerned with scientific plausibility, up to a certain point. *Plausibility* is the right word here, as it's not the same as *likelihood*, much less *prediction*.

The main thing I stretched in terms of scientific plausibility, or rather compressed, was my timeline. There was a scientist named Paul Birch who wrote two papers that were very helpful to my thinking, called 'Terraforming Mars Quickly' and 'Terraforming Venus Quickly'.[3] For Mars, he postulated that it could be terraformed in fifty years, by using all possible technologies available to us. Other scientists working in the 1980s suggested it would take more like 500 to 50,000 years, depending on methods used. So, to make my novel work the way I wanted it to, I chose quite a rapid timeline, and showed Mars being terraformed in about 200 years. That, plus a longevity treatment lengthening human lifetimes, gave me a novel in which the planetary action could happen during the lifetimes of a single set of characters, who therefore saw the whole thing.

I got interested in this project in the late 1970s, after the data from NASA's Viking probes came back to Earth, was digested by the scientists, and published in forms that laypeople could understand. A great *Atlas of Mars* was published by NASA,[4] along with a geological map, and Michael H Carr of the United States Geological Survey published his wonderful first analysis of all this data, called *The Surface of Mars*.[5] Then, also, a group of graduate students at the University of Colorado, who called themselves the Mars Underground, convened a series of 'Case for Mars' conferences, and the American Astronomical Society published their proceedings as *The Case for Mars,* in four fat volumes.[6] These proceedings were really design fiction: they went into detail on matters

of a Martian surface suit, for instance, including how the gloves would attach to the suit. They also discussed the early habitats, which were to be built in part by using local Martian materials – brick, strengthened by shredding the first giant parachutes and adding their nylon to the brick matrix. Also, lots of bamboo – which grows fast and is strong – used as a building material for interiors, and so on. They covered every aspect that design covers here on Earth.

I read and made notes on all these books, and met with or called many of the scientists involved to ask them questions. Right after *Red Mars* came out, the University of Arizona published a thick book of papers called *Mars*,[7] which contained everything we knew about Mars up to about 1990. That gave me things to put in *Green* and *Blue Mars* that I hadn't known when writing *Red*, which was very nice – it looked as though I had been saving neat information, when really I learned it as I went along. The writing took about seven years, so there was time for new information to be compiled and analysed.

The cities in my book I mostly designed myself using principles from Christopher Alexander's *A Pattern Language*,[8] from Paolo Soleri, and from architect Paul Sattelmeier, who designed the pre-fab 'disk house' seen in *Blue Mars*. I had also visited the classic Greek archaeological sites on Crete, which accounts for the Minoan strand in my Martians.

I was working at a time where there was a lot of new scientific information about Mars, and a lot of speculative utopian design from the 1970s' ecotopian non-fiction literature, and all of it combined very nicely. And yet at the same time there wasn't too much of it, as maybe there is now. I felt like I could come close to knowing everything about Mars that there was to know. The books containing that information filled only a couple of bookshelves. It wouldn't be that way now.

More than twenty years on from the trilogy, we now live in a world where space entrepreneurs have overtaken governments as the primary drivers of space exploration. Elon Musk talks of putting people on Mars within a handful of years. Not since the 1950s have film and television been so taken with Mars. Plausibly or not, it is once again part of the zeitgeist. How does this affect how you look back on the trilogy? And do you see this playing out as it did in the books, as a battle between science and commercial exploitation?

Maybe. From my perspective, interest in Mars has come in waves that were tied to the robotic landers we sent there. There was a huge surge of interest in the late 1990s when NASA announced they thought they had seen signs of fossil Martian bacterial life. Even more so when Sojourner landed and travelled around, and, later, Spirit and Opportunity. So the current interest is not really coming out of any long absence of interest but is really just the latest wave in a century and a half of almost continuous interest. One thing that may have caused a lull in this interest was the 9/11 attacks, which put us into a new moment of history that made interest in other planets go away for a while. But then it came back.

When I look back to my trilogy, I see past it to the previous waves. First, Percival Lowell and the idea that there might be intelligent Martians building canals in their own version of catastrophic climate change. That led to HG Wells, Bogdanov in Russia, Lasswitz in Germany, Edgar Rice Burroughs in the United States, and so on. After that vision dissipated, because of the scientific information that Mars has no atmosphere to speak of, we got the 'dry Mars' of Bradbury, Clarke and Philip K Dick. Then Viking, which was my moment.

Now, if we look at all those waves, I think the one we're experiencing now is the same as before – a mix of science and science fiction, with the fiction including, just as with Percival Lowell, a fair amount of hype, and fundraising disconnected from reality.

For me, the biggest changes are the new things we've learned, principally these: there might be bacterial life still alive there underground. Also, there may be a lot less nitrogen than we thought there was; and terraforming Mars needs nitrogen. Lastly, there are perchlorates on the surface in quantity, and these salts are poisonous to humans. They may present a huge problem to humans living there. So these are the main things that would have changed my trilogy if I had known about them at the time I wrote – if science had known about them, I mean.

Elon Musk will not be putting people on Mars within a handful of years. He is a great technical and business leader, doing wonderful work with cars and batteries and solar paint and rocketry – all very important work. Mars is more his hobby, as far as I can tell, and his pronouncements on that topic have a kind of boosterism that is not so evident in his more earthly projects. That's OK with me: I like his hobby. But Mars will continue to present us with a really difficult landing problem. We still only have a fifty per cent success rate for landing robotic landers. We'll want better for human landings. So the financial investments needed are really big, and there is never any economic payoff. It's the same problem on the Moon, only

made more extreme by sheer distance and by the technical challenges of landing on Mars.

When you ask about commercial exploitation of Mars – no. There will be no such thing. There's nothing on Mars we can go get and bring back that we don't already have here in more concentrated forms. Really it's about scientific research, and the robotic explorers are doing an excellent job of that. No one will pay to put people on Mars, because there is no profit in it.

Given that situation, NASA will likely team up with other national space programmes someday, and land scientists on Mars. At that point it will be much like Antarctica has been since about 1956, which is to say, a research site where scientists and support workers arrive, work for a couple of years, and then return to Earth. As such, it will be very interesting! But it won't be transformed, nor transformative.

There was a planned Endemol reality-TV show (*Mars One*, which never aired) that invited participants to become the first colonisers. More than 200,000 people applied. What do you think the appeal is?

I think there is a kind of category error going on here, where people are thinking, I don't like this world, so I would be better off on another world. That's wrong. Mars is poisonous to humans in multiple ways, and life there would be even more confined than life on Earth in late capitalism. The real change, if you wanted to make one, would be to alter human life on Earth to a system more humane and rewarding. To the extent that this looks impossible, people dream of escape. It's not wrong to want things to be different and better – that's every hope, really, and for sure it's the utopian hope for a better society. But that can only be made to happen here on Earth.

It's also true that people want meaning. We all want that, and for sure living on Mars would make for an automatic meaning to life – you'd be doing something hard and interesting, new and meaningful. But again, the thing to do is to direct that desire to something that you can really do. Which means altering life on Earth.

Most people would assume that life on Mars would be high-tech and futuristic, whereas it will more likely be a world of ad hoc jury-rigged design solutions. In some ways, it would be quite primitive. But survival at inhospitable frontiers is the ultimate design fantasy – a place to make the world anew. Will Mars be the American West with EVAs [Extravehicular Activity suits], or is it different this time?

It's definitely not the American West with EVAs. This is thinking by historical analogy, in a situation that in fact would be entirely new, and nothing like the American West in the nineteenth century. That analogy isn't really anything but a fantasy of escape.

If there is any historical analogy that works for Mars, it would be Antarctica. Americans have built a little town on Ross Island called McMurdo, starting in 1956 and working on it continuously until now. It has the kind of jury-rigged design you reference, sometimes called 'adhocitecture', where the old plywood buildings from the beginning are slowly added to over the decades, so that now there are antique wooden structures next to high-tech composite laboratories.

On Mars, we'll probably first land a number of robotic landers that hold all the elements needed for human habitation. When people finally arrive, all this stuff will already be there, waiting for assemblage and inhabitation. Then it would make sense just to keep going to that first spot, as we have kept going to McMurdo in Antarctica, because there's already an infrastructure there. With each decade, improvements in materials and technologies and designs will mean that the new stuff doesn't match the old stuff, but the old and the new will coexist because people will always put to use what's already at hand.

I guess I too am using a historical analogy to think about what is going to actually be a very new situation. But I think mine is more accurate for the situation facing us on Mars.

Much of the work in the exhibition is speculative. We talk of 'design fiction' rather than science fiction – an attempt to give highly speculative situations specific forms and tools. What do you think might be the advantages of this approach?

I think design fiction is a real genre, useful as it brings the focus onto how much design influences the way we live. I live in a designed community, for instance – Village Homes, Davis, California. It was designed to see if a village like those in Europe could be recreated in an American suburban context, as a replacement for, and improvement on, the standard post-war suburbia. After almost thirty years of living here, I can testify that design changes lives, that the built infrastructure shapes habits by creating some opportunities, or by cancelling others out, depending. So design is, I think, about the lived reality of speculative utopian thought. It makes concrete what could remain rather fuzzy about how to make a better, more meaningful life.

By doing this kind of thinking about the hypothetical but possible situation on Mars, we expose the values that we're hoping to live by here on Earth. So it's great to think about design as a kind of grammar for living in the material world, any material world.

The idea of terraforming Mars could be seen as the total design project: it is redesigning the atmosphere. This is obviously highly contentious, as it is in your trilogy. One of the contributors to the *Moving to Mars* exhibition, a speculative designer called Daisy Ginsberg, proposes colonising Mars with bacteria and flora instead of humans, and letting an ecosystem evolve over thousands of years. So, in other words, treating Mars as a lifeboat for plants not people. Does this solve any of our ethical dilemmas or is it just as invasive?

This would be what in the 1980s Chris McKay and others began referring to as 'ecopoiesis' – it's mentioned in my Mars books as such.

McKay was one of the original Mars Underground members, and he's been working all along on Mars and many other planetary projects at NASA's Ames Research Center. He's helped me with many of my books, and taught me a lot by his example about how a working scientist can make a difference in the public discussions about his field.

His argument was that Mars might have had bacterial life back in its early warm, wet period, before it experienced a climatic crash so complete that it lost most of its atmosphere. That being the case, if we reintroduced Terran bacteria on Mars and left them alone (or added heat to the surface by way of space mirrors and so on), we would have an experiment in planetary life and evolution that would be more interesting than anything else we could do here. If life took root and prospered there, changing the atmosphere the way our archaea added oxygen to the early Earth, this would be a kind of 'landscape restoration', as well as an experiment that might teach us a lot about maintaining life on Earth in a good balance. It is this that McKay called ecopoiesis, although the word originated with Robert Haynes.[9] He suggested once that a big transformation of Mars, its terraformation by way of life introduced in this slow way, might take about a hundred thousand years. 'But what's the hurry?', he would add.

One of the characters in your Mars trilogy, Arkady Bogdanov, says: 'We must terraform not only Mars, but ourselves'.[10] He meant politically, imagining a new, egalitarian society. What other ways might we need to change ourselves if we are really to become an interplanetary species? Or what ways might Mars simply change us?

I don't think we are destined to become an interplanetary species. Or, if we do that, it would only be because we have managed to get ourselves into a sustainable balance with the rest of the biosphere here on Earth.

It's possible that because we co-evolved with Earth, we will never be healthy anywhere else. If Mars turns out to be habitable and terraformable, that might be a project that takes thousands of years. But if we don't kill ourselves off, there will be thousands of years to come for humanity. It's just that we need to make it work here, now, for it to be at all possible anywhere off this planet.

I can imagine us getting into a balance here on Earth, and then establishing bases on

a lot of other bodies in the solar system, similar to our Antarctic bases now – places interesting and useful, but not at all central or transformative to history. Then, if all that happens and goes well, the Mars terraforming project might be tried, and might even succeed. At that point, we really would have two inhabited planets, but that might be 3,000 or 10,000 years from now, which would be fine.

By that point, Mars would have had to change us in the following ways: we will have to have learned that Earth is always the one and only centre of the human story, and our only long-term viable home. And we will have learned patience, and long-term historical thinking. The Martian perspective is indeed a bit alien: it generates the 'estrangement-effect' that Brecht used to talk about when designing his theatre experiences.[11] It's there to make our own life on Earth seem as strange and tenuous to us as it really is. Once we get that angle of vision, we might be more careful with Earth. I hope so.

1   Kim Stanley Robinson, *Red Mars* (London: HarperCollins, 1992); Kim Stanley Robinson, *Green Mars* (London: HarperCollins, 1993); Kim Stanley Robinson, *Blue Mars* (HarperCollins, 1996).

2   Kim Stanley Robinson, *Red Mars* (London: HarperCollins, 1992), 402.

3   Paul Birch, 'Terraforming Mars Quickly', *Journal of the British Interplanetary Society*, 45 (1992), 331-40; Paul Birch, 'Terraforming Venus Quickly', *Journal of the British Interplanetary Society*, 44 (1991), 156-67.

4   Astrogeology Science Center, *Atlas of Mars: the 1:5,000,000 map series* (Washington, DC: National Aeronautics and Space Administration, 1979).

5   Michael H Carr, *The Surface of Mars* (New Haven, CT: Yale University Press, 1981).

6   Penelope J Boston (ed.), *The Case for Mars: proceedings of a conference held April 29-May 2, 1981, at University of Colorado, Boulder, Colorado* (San Diego, CA: Univelt for the American Astronautical Society, 1984); Christopher P McKay (ed.), *The Case for Mars II: proceedings of the second Case for Mars Conference held July 10-14, 1984, at the University of Colorado* (San Diego, CA: Univelt for the American Astronautical

Society, 1985); Carol R Stoker (ed.), *The Case for Mars III: Proceedings of the third Case for Mars Conference held July 18-22, 1987, at the University of Colorado, Boulder, Colorado* (San Diego, CA: Univelt for the American Astronautical Society, 1989); Thomas R Meyer (ed.), *The Case for Mars IV: the international exploration of Mars - mission strategy* (San Diego, CA: Univelt for the American Astronautical Society, 1997).

7   Hugh H Kieffer et al. (eds.), *Mars* (Tucson: University of Arizona Press, 1992).

8   Christopher Alexander et al., *A Pattern Language: Towns, Buildings, Construction* (New York: Oxford University Press, 1977).

9   Robert H Haynes, 'Ecco Ecopoiesis: Playing God on Mars', in *Moral Expertise: Studies in Practical and Professional Ethics*, ed. Don MacNiven (London: Routledge, 1990), 161-83.

10 Kim Stanley Robinson, *Red Mars* (London: HarperCollins, 1992), 113.

11 See Bertolt Brecht, *Brecht on Theatre* (3rd edn), eds. Marc Silberman, Steve Giles and Tom Kuhn (London: Bloomsbury Methuen Drama, 2015), 143.

Rick Guidice, *The Torus Wheel*, 1976. Proposed during the 1975 Stanford summer school, the Stanford Torus is a ring-shaped space station that would rotate to induce artificial gravity through centrifugal force.

Donald E Davis, *Stanford Torus, interior view*, 1975.

In the summer of 1975, NASA Ames Research Center and Stanford University held a ten-week workshop to design free-floating space settlements. The programme was unique in bringing together professionals and students from a wide pool of disciplines, including engineers, physicists, architects and social scientists, led by Gerard O'Neill of Princeton University. The settlements devised during the workshop – including the Bernal Sphere, Stanford Torus and O'Neill Cylinder – were richly illustrated by prominent space artists Rick Guidice, Don Davis and Syd Mead.

Rick Guidice, *Stanford Torus, interior view*, 1975.

Rick Guidice, *Agricultural modules in a cutaway view*, 1978.

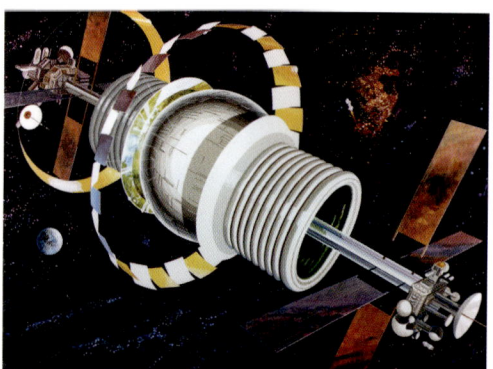

Rick Guidice, *Bernal Sphere*, 1976.
A spherical habitat that rotates to
generate Earth-like levels of gravity
through centrifugal force. Although
originally developed by John Desmond
Bernal in 1929, Gerard O'Neill proposed
a modified version at the 1975 Stanford
summer school.

Rick Guidice, *Cutaway view of Bernal Sphere*,
1976.

Donald E Davis, *Bernal Sphere construction*, 1976.

Rick Guidice, *Bernal Sphere, interior view*, 1976.

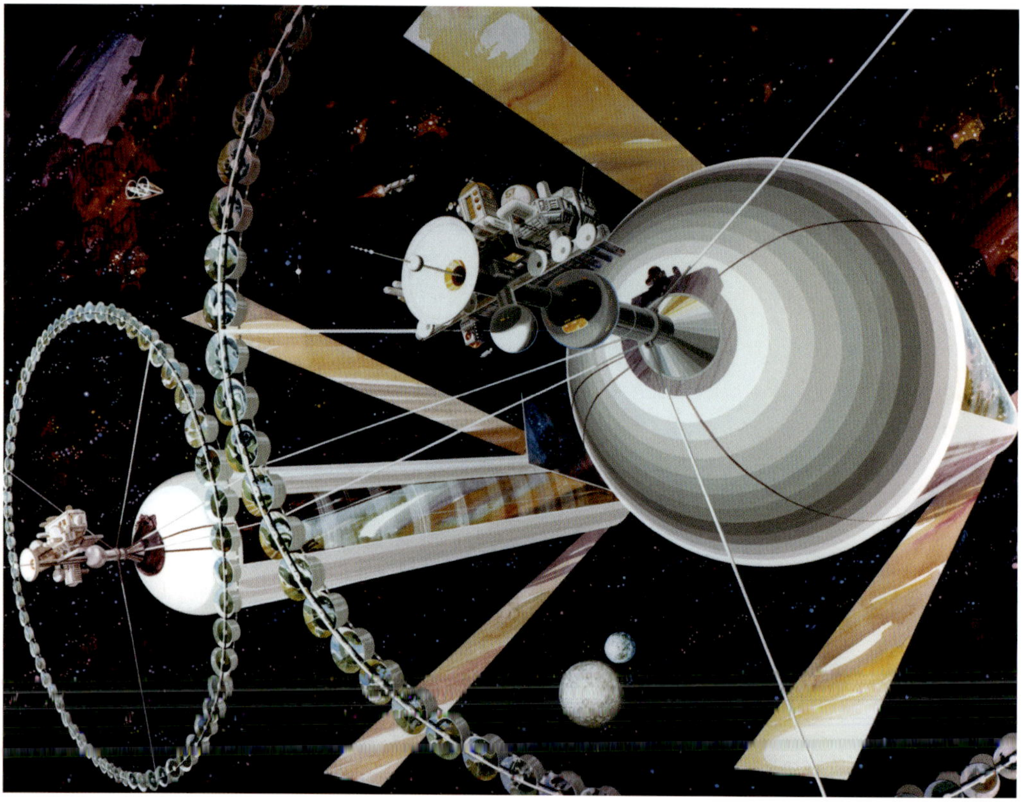

Rick Guidice, *A Pair of O'Neill Cylinders, exterior view*, 1975. Cylindrical space colonies that would rotate to generate gravity by centrifugal force and be deployed in pairs to offset destabilising gyroscopic effects were proposed by Gerald K O'Neill at the 1975 Stanford summer school.

Donald E Davis, *O'Neill Cylinder, interior view*, 1975.

Rick Guidice, *O'Neill Cylinder, interior view*, 1975.

Bjarke Ingels Group, Mars Science City exterior views, renders, 2017.

Designed by Danish architecture firm Bjarke Ingels Group, Mars Science City is a Mars-simulation complex currently in development in the desert outside Dubai. Covering an area of 180,000 square metres (1.9 million square feet), the complex will include laboratories, a museum and a space for researchers to live for up to a year.

Bjarke Ingels Group, Mars Science City interior views, renders, 2017. Clockwise from top: permanent exhibition space; office; underwater museum; canyon; aquaponic farm.

SpaceX, Mars settlement with Starship, render, 2018.

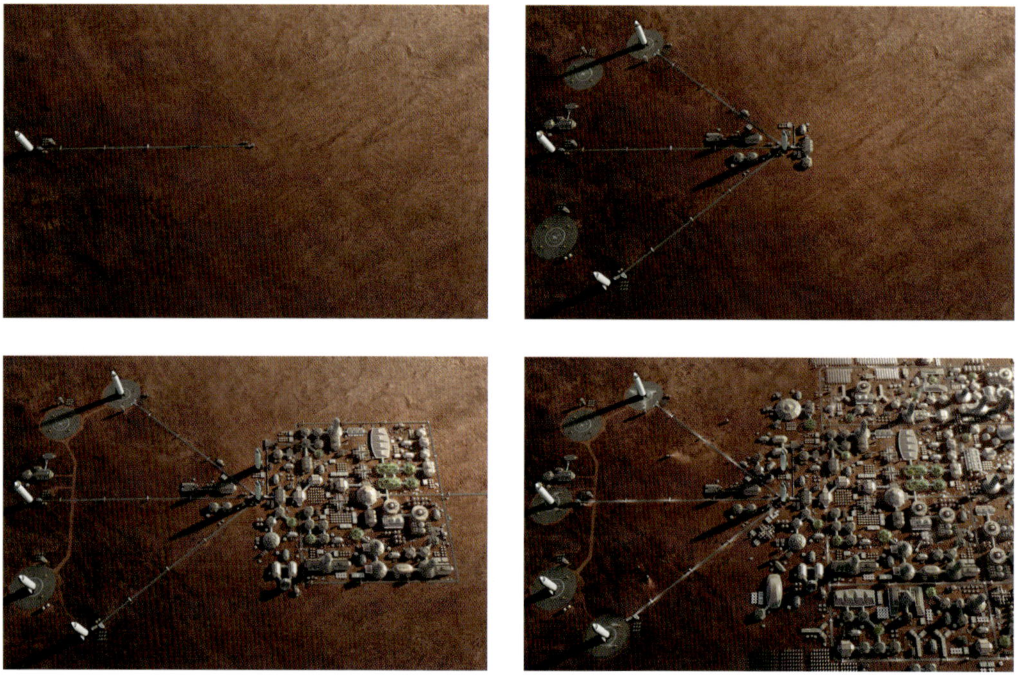

SpaceX, the development of a Mars city, film
stills, 2018.

SpaceX, ilustration showing the changing surface of Mars as it is terraformed, 2017.

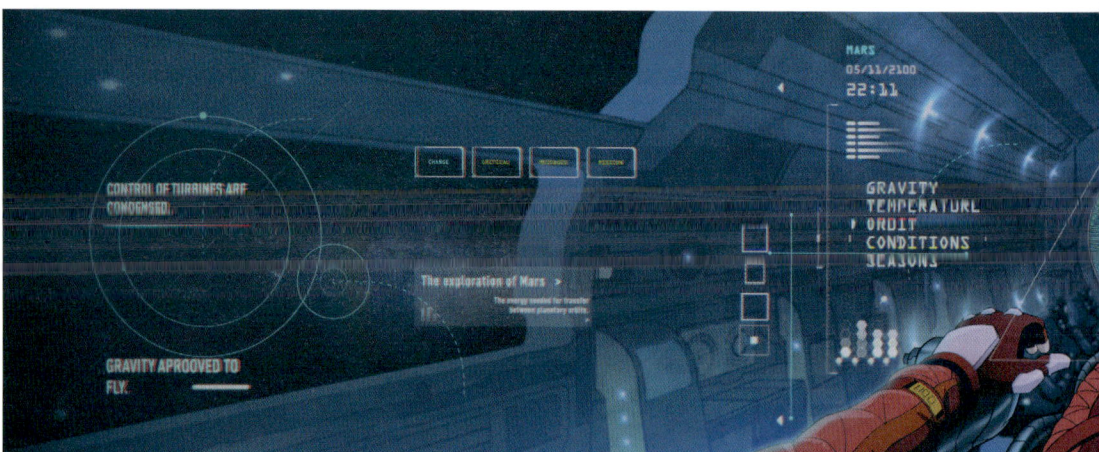

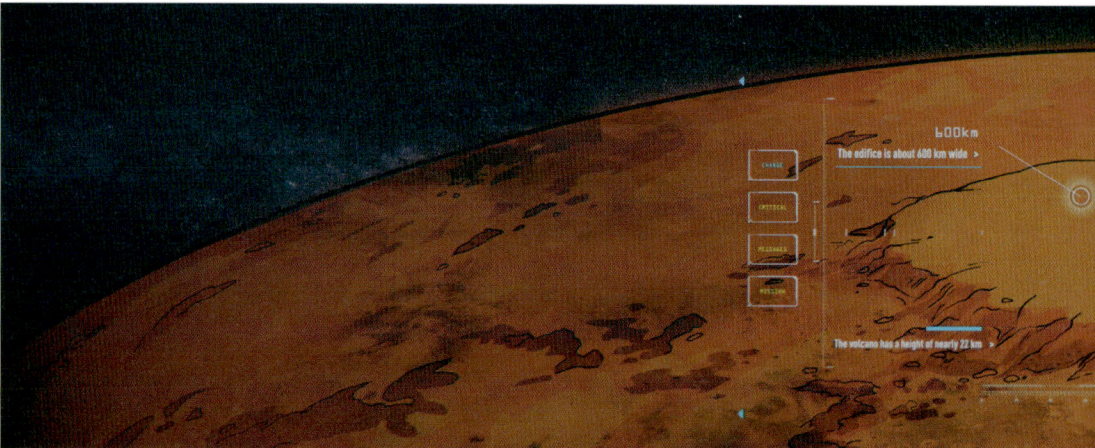

Creative Directors and Executive Producers: Thomas Ermacora, Liana Brazil and Russ Rive, *Mars 2100*, film stills, 2019. *Mars 2100* is an immersive animated experience created and designed by Thomas Ermacora and SuperUber, and premiered at the Design Museum. Looking to the near future of Mars exploration, the film takes visitors on the emotional journey of future astronauts and explorers as they prepare to leave Earth and dedicate their lives to scientific research on the Red Planet.

Script Writer: Alexandre Rossi   Director of Operations: Fabiano Martins
Lead Animator: Bernardo Leite   Director of Technology: Carlos Oliveira
Illustrator: Andres Ramos   Producer: Mariana Ferman
Animators: Rodrigo Borges, Rodrigo Mantega Project Manager: Ilana Markus
Sound Design and Music: Nick Ryan Studio   Technical Sound Design: Davey Williamson
Voice Narrators: Trine Garrett, John Kapansa Technical coordination: Blue Elephant

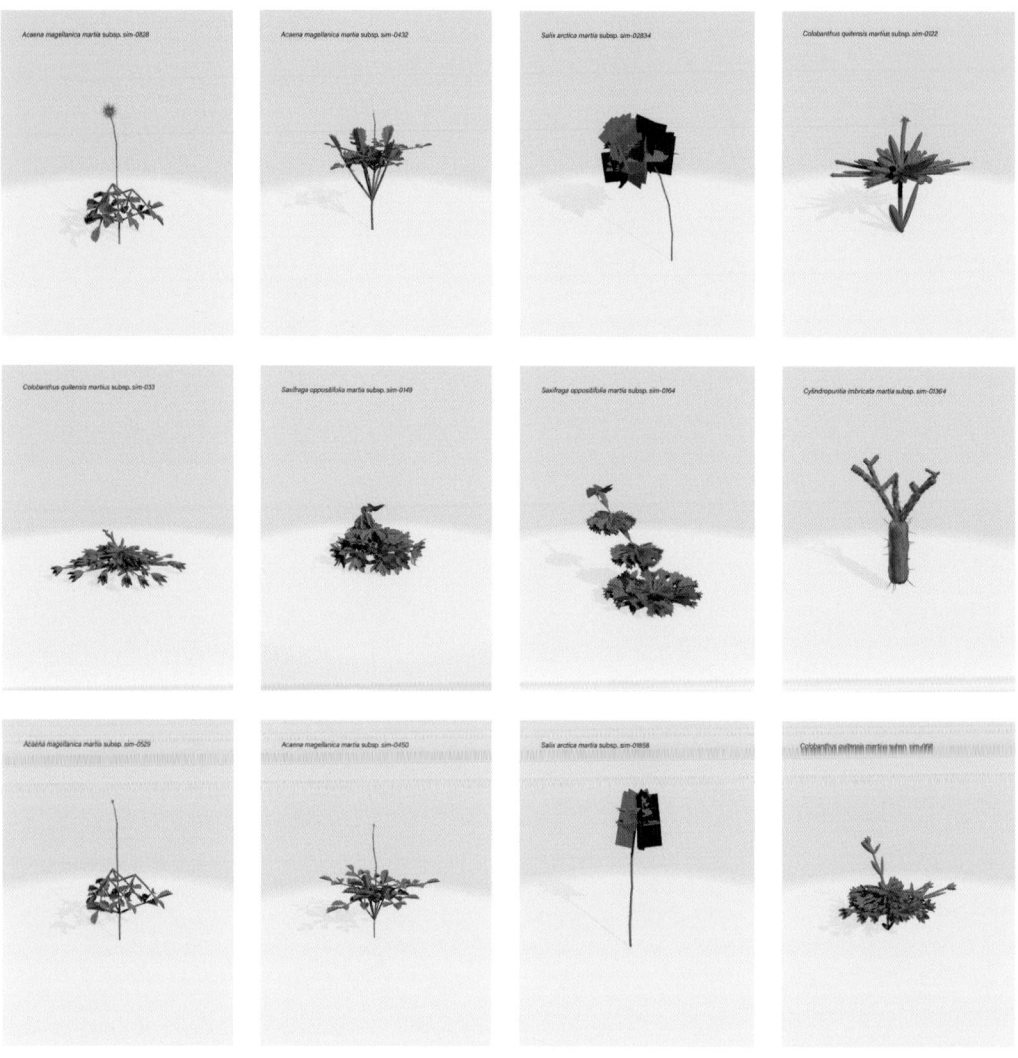

Alexandra Daisy Ginsberg, pioneers and selected generated subspecies from *The Wilding of Mars*, 2019.

Commissioned by the Vitra Design Museum and the Design Museum in London, *The Wilding of Mars* is both a thought experiment and an artwork. It explores an alternative future in which we might colonise Mars with bacteria and plants, but do not settle on the planet ourselves. Multidisciplinary artist Alexandra Daisy Ginsberg researched the types of plant life that might survive on Mars, starting with cyanobacteria to absorb light and heat up the icy surface.

The evolution of the planet – together with the new life forms that might evolve – is explored in a computer simulation that runs at a speed showing a million years in an hour. As the planet's biosphere changes, new species are introduced, and together these evolve and change in unforeseeable ways. These growth patterns are generated by a gaming engine, so every time you see this work the new biosphere will be slightly different.

Alexandra Daisy Ginsberg, pioneer species *Cylindropuntia imbricata* from *The Wilding of Mars*,
2019.

THE WILDING OF MARS
Alexandra Daisy Ginsberg

A treacherous new frontier for humankind to conquer; an empty terrain, rich with opportunity and resources for humanity to exploit; a backup planet for humans to flee to from our dying Earth; a pitstop for our evolution into a multiplanetary species. Mars – an alien globe spinning between 55 million and 400 million kilometres (34 million and 250 million miles) from Earth – is also a utopia where humanity will progress into a more civilised society. If anything, these popular contemporary imaginings of Mars expose a single idea about humans: *Homo sapiens* is a solitary species that can emancipate itself from Earth and from the roiling symbiotic nature that we evolved with and flourish among. In colonising Mars, enthusiasts promise, we will become *better* people. If Mars is accessible to most humans in any tangible way, it is as a revealing place to reflect back on Earth, and on human nature.

NASA, 'The Blue Marble', Earth photographed from space by the crew on board Apollo 17, 1972.

Humans saw Mars in its planetary fullness long before we ever saw our own. Its orange disc has been scrutinised by astronomers for centuries, becoming a place for humans to project what we know about our world onto another. For scientists, studying another planet gives insight into our own, from its geology to the emergence of life. In 1972, when a photograph of Earth seen from space was widely published for the first time, humanity finally had the privilege of a holistic, global view of our home.[1] But the only way we can really see Mars is through the lens of our earthly understanding. Examining that perspective enables us to look more closely at our own actions.

For fifty years, and well into the last century, Mars was seen as the 'dying planet'.[2] At the turn of the twentieth century, American businessman and astronomer Percival Lowell famously misinterpreted Italian astronomer Giovanni Schiaparelli's false observations of *canali* – a global network of channels – as a lost civilisation's desperate attempt to build a canal system to irrigate an overheating planet.[3] Concerns about planetary limits were already emerging on Lowell's Earth as nineteenth-century industrialisation increased pollution, the exploitation of resources and desertification. Lowell warned of Earth 'going the way of Mars'.[4]

While growing scientific understanding of Mars has downsized our hopes of encountering civilised – or perhaps any – life on Mars, other dreams have prevailed, including that of 'terraforming'. Transforming another planet's conditions to make it more like Earth, this hypothetical process would allow life from Earth to survive elsewhere. Today, advocates are clamouring to terraform Mars, citing climate breakdown and biodiversity decimation here, sidestepping human responsibility for these crises. Techno-utopian Elon Musk laments humans 'stuck' on Earth and prey to some extinction event, chivvying us towards a multiplanetary destiny, accelerated by his private space-exploration business.[5] If Earth is going the way of Mars, then, for many would-be terraformers, Mars *must* go the way of Earth. Mars will be the venue for our technological and biological resurrection. It is a solutionist proxy for concerns about human impact on Earth: making Mars liveable would prove that we can

Alexandra Daisy Ginsberg, work in progress
from *The Wilding of Mars*, film still, 2019.

Alexandra Daisy Ginsberg, work in progress
from *The Wilding of Mars*, film still, 2019.

Alexandra Daisy Ginsberg, work in progress
from *The Wilding of Mars*, film still, 2019.

resurrect life here on Earth without changing our behaviour.

I was asked to make a new work for the *Moving to Mars* exhibition that addressed my critical view of Mars colonisation, but still operated within the frame of design, the optimistic process by which humans turn existing conditions into preferable ones.[6] If optimism means imagining that this is the best of all possible worlds, that world for me is one where we acknowledge the limits of the human-centric view of life. Can we imagine humans not only as exploiters, but as protectors of other species? Life could flourish beyond Earth, but it does not have to include us. Instead of cultivating Mars with 'useful' crops, I wanted to reveal the pressures that humans place on nature as we exploit it for our own benefit. I decided to seed a planetary wilderness, where other species could thrive.

The result, *The Wilding of Mars*, is shown as two simulations that run in parallel, built in the computer-game engine Unity. Visitors watch Mars being colonised over a million years, viewed in a human hour. But the colonisers are cyanobacteria, fungi, lichens and plants: there are no humans. If the wild is a 'state of existing in relative freedom from human interventions', this is a wilderness that humans may create, but never visit.[7] The planet is seeded with extremophile Earth life forms that might tolerate the harsh conditions of Mars. The premise is still inescapably colonial, and so it acknowledges colonial language: we watch the sixteen 'pioneer' species spreading north from Mars' South Pole. The invasive species arrive in stages as they make the environment increasingly tolerable, contributing to a new ecosystem. Strange new subspecies evolve that are better adapted to Mars; harsh seasons kill off some. All are documented. Life takes Mars in a different direction, and Mars may take life elsewhere. The aim is not to terraform Mars and make it Earth-like; it is simply a repository for the mechanism of life.

The Wilding of Mars prioritises a non-human perspective: we can watch plants growing and spreading across the landscape, while voyeuristic camera angles heighten the sense of human intrusion and alienation. Two versions of the simulation run at the same time to show that many possible worlds can emerge, challenging the assumption that the only outcome of space colonisation is human benefit.

The work confronts the disturbing vision that Mars could be comfortably habitable, a place we could exist, or a world we should colonise. Far from the painterly renders of humans busy on Mars, a life removed from Earth's evolved, natural life support would in reality be terrible: imagine constantly protecting your delicate DNA, flesh, and membranes from devastating temperatures, radiation, poisonous air, bone-deteriorating gravity and knife-sharp dust. Imagine your only companions being other trapped humans, machines and utilitarian crops like potatoes. I find this devastating, not progressive. Now imagine a million humans in these harsh colonies

Alexandra Daisy Ginsberg, work in progress from *The Wilding of Mars*, film still, 2019.

by 2060, as Musk proposes.[8] Think about the social hierarchies, the strip-mining and even the prisons necessary to manage the human animal under such threatening conditions. All so we can escape a planet that we ransacked over a few centuries. We won't have long to change our expectations of life.

While the technical barriers – if surmountable – of getting so many people 'off planet' would probably further devastate Earth, the cultural rhetoric and economic ambition should cause greater alarm. Mars is breezily described by would-be colonists as a wild frontier rich for exploration and exploitation. They make no reference to the terrible social and environmental legacy of colonial actions on Earth – imperialism, genocide, the displacement and erasure of indigenous cultures and rights, economic inequity, the destruction of biodiversity and resource exploitation – all for the benefit of colonisers.[9] Just as the frontiers that colonisers 'discovered' on Earth were not actually empty of people, organisms or ecosystems, Mars is not necessarily a virgin frontier that a few colonisers have a right to. In the postcolonial era, any suggestion that repeating colonialism is harmless should be scrutinised for other interests.

Limited technological surveys of Mars have not yet revealed life there, but the planet may have had, or could still have, its own indigenous life forms. In 1967, signatory states of the United Nations' International Outer Space Treaty agreed to avoid 'harmful contamination' of other planets.[10] Of course, humans are not the only biological colonisers: other animals, plants, fungi or bacteria colonise the world around them too. Scores of other organisms – including the bacteria that outnumber our own cells, and which our bodies rely on for survival – would travel with us. As we populate Mars with this menagerie of Earth life, indigenous life may be destroyed. The treaty demands that 'the exploration and use of outer space should be carried on for the benefit of all peoples' but makes no mention of non-human rights. Astronomer Carl Sagan sagely wrote, 'If there is life on Mars, I believe we should do nothing with Mars. Mars then belongs to the Martians, even if the Martians are only microbes … the preservation of that life must, I think, supersede any other possible use of Mars.'[11]

Today, the treaty is also under threat as private interests such as SpaceX, which may not be within their legal scope, race to be the first to Mars. Habitability is a barrier but so too is cost. As corporations with the necessary capital strive to privatise the 'commons' of space, they replicate the structures of imperial and capitalist colonialism. Who benefits from their colonies and access to resources, here or on Mars? Who will go and who will remain on Earth? And who, or what, will be exploited on the way? Creating Mars as a better world for all, whether human or non-human, is unlikely. This utopia will most likely mirror the social and environmental injustices of extractive capitalism on Earth, where the few decide what is better for themselves, while exploiting worlds around them.

How can we think differently about planetary futures? Instead of colonising Mars, we should – as Mark McCaughrean, an advisor at the European Space Agency, tweeted in response to Musk's plans – first focus on 'making the Earth a sustainable multi-species planet'.[12] However, a planetary perspective is useful to help us see the limits of human imagination. Mars is useful, but not as a utopia. It is a 'heterotopia', a place that is neither better nor worse, but different. It is another world from which we can observe our own nature.[13]

Life can take other paths. *The Wilding of Mars* is a heterotopia that creates what space anthropologist Lisa Messeri calls a 'double exposure', a way to simultaneously experience both Earth and Mars, future and present.[14] It invites us to reimagine Mars not only as a place for ourselves. Might giving the planet to other life forms be the ultimate unnatural act for humans?

1   Apollo 17 Crew, 'AS17-148-22727', *NASA*
    (7 December 1972), https://spaceflight.
    nasa.gov/gallery/images/apollo/apollo17/
    html/as17-148-22727.html [Accessed
    6 June 2019]. See 'Earthrise; or, the
    Globalization of the World Picture',
    American Historical Review (June 2011),
    602-30.

2   Robert Markley, *Dying Planet: Mars in
    Science and the Imagination* (Durham, NC:
    Duke University Press, 2005), 2.

3   See K. Maria and D. Lane, *Geographies
    of Mars: Seeing and Knowing the Red
    Planet* (Chicago: University of Chicago
    Press, 2010).

4   Percival Lowell, *Mars as the Abode of
    Life* (New York: Macmillan, 1909), 122.

5   Nadia Drake, 'Elon Musk: A Million
    Humans Could Live on Mars By the 2060s',
    National Geographic (27 September
    2019), https://news.nationalgeographic.
    com/2016/09/elon-musk-spacex-exploring-
    mars-planets-space-science/ [Accessed
    6 June 2019].

6   Herbert A Simon, 'The Science of Design:
    Creating the Artificial', *Design Issues*,
    4/1-2 (1988), 67-82.

7   Bradley Cantrell, Laura J Martin and
    Erle C Ellis, 'Designing Autonomy:
    Opportunities for New Wildness in the
    Anthropocene', *Trends in Ecology &
    Evolution*, 32/3 (2017), 156-66, https://
    www.cell.com/trends/ecology-evolution/
    fulltext/S0169-5347(16)30237-3 [Accessed
    9 September 2019].

8   Ibid.

9   Ryan F Mandelbaum, 'Decolonizing Mars:
    Are We Thinking About Space Exploration
    All Wrong?', *Gizmodo* (20 November 2018),
    https://gizmodo.com/decolonizing-mars-
    are-we-thinking-about-space-explorat-
    1830348568 [Accessed 6 June 2019].
    See also https://www.decolonizemars.org/
    [Accessed 9 September 2019].

10 United Nations Office for Disarmament
    Affairs, *Treaty on Principles Governing
    the Activities of States in the
    Exploration and Use of Outer Space,
    including the Moon and Other Celestial
    Bodies* (London, Moscow and Washington,
    DC: United Nations, 1967), http://
    disarmament.un.org/treaties/t/outer_space
    [Accessed 6 June 2019].

11 Carl Sagan, *Cosmos* (New York: Random
    House, 1980), 108.

12 Hannah Devlin, 'Life on Mars: Elon Musk
    reveals details of his colonisation
    vision', *Guardian*, 16 June 2017, https://
    www.theguardian.com/science/2017/jun/16/
    life-on-mars-elon-musk-reveals-details-
    of-his-colonisation-vision [Accessed
    6 June 2019].

13 Michel Foucault, 'Of Other Spaces,
    Heterotopias' (Des Espace autres, 1967,
    trans. Jay Miskowiec), *Architecture,
    Mouvement, Continuité* 5 (1984), 46-9,
    https://foucault.info/doc/documents/
    heterotopia/foucault-heterotopia-en-html
    [Accessed 6 June 2019].

14 Lisa Messeri, *Placing Outer Space:
    An Earthly Ethnography of Other
    Worlds* (Durham, NC: Duke University
    Press, 2016), 30.

END MATTER

BIOGRAPHIES

EDITORS

JUSTIN MCGUIRK is chief curator at the Design Museum. He has been the design critic of the *Guardian*, the editor of *Icon* magazine and the Head of Design Curating and Writing at Design Academy Eindhoven. He is the author of *Radical Cities: Across Latin America in Search of a New Architecture* (2014), and co-editor of *Fear and Love: Reactions to a Complex World* (2016), *California: Designing Freedom* (2017) and *Home Futures: Living in Yesterday's Tomorrow* (2018).

ANDREW NAHUM is a curator and author on the history of technology and design. He has created numerous exhibitions at the Science Museum and recently curated the acclaimed *Ferrari: Under the Skin* at the Design Museum (2017–2018). His books include *Fifty cars that changed the world,* (2009, 2017), *Alec Issigonis and the Mini* (2004), *Frank Whittle: invention of the jet* (2017) and *Ferrari: Under the Skin* (2017).

ELEANOR WATSON is assistant curator at the Design Museum. Her previous exhibitions include *Beazley Designs of the Year* (2017 and 2018) and *Imagine Moscow: Architecture, Propaganda Revolution* (2017). Eleanor is responsible for the museum's programme of free public displays, including the first retrospective of UK architect Peter Barber, '100 Mile City and Other Stories'. Outside of her work at the museum, she is acting as curator of the 2019 edition of *Global Grad Show*.

CONTRIBUTORS

MIKE ASHLEY has been researching and writing about science fiction and fantasy for over fifty years and has compiled over a hundred books including *The Age of the Storytellers* (2006), *Out of This World* (1989) and the multi-volume *History of the Science Fiction Magazine* (1974). He was awarded the Pilgrim Award for Lifetime Achievement in science fiction research in 2002.

ALEXANDRA DAISY GINSBERG is a multidisciplinary artist examining human relationships with nature and technology. She is lead author of *Synthetic Aesthetics: Investigating Synthetic Biology's Designs on Nature* (2014). *Better*, her PhD from the Royal College of Art, interrogated how powerful dreams of 'better' futures shape the things we design.

LYDIA KALLIPOLITI is an architect, engineer and scholar living in New York. She is assistant professor at the Cooper Union, founder of ANAcycle thinktank and author of *Closed Worlds, Or, What is the Power of Shit* (2018). She holds a SMArchS from MIT and a PhD from Princeton University.

STEPHEN L PETRANEK is the author of *How We Will Live on Mars* (2015). He has spoken on the TED conference main stage three times, and his TED talk about Mars has been watched nearly five million times. He is co-executive producer of National Geographic's *Mars* documentary series. He was previously editor-in-chief of *Discover* and *The Washington Post Magazine* and editor for science at *Life Magazine*.

KIM STANLEY ROBINSON is an acclaimed writer of science fiction who has written more than twenty novels including the bestselling Mars trilogy *Red Mars* (1992), *Green Mars* (1993) and *Blue Mars* (1996). He has won every major award for science fiction writing including the Nebula, Hugo, Asimov, John W. Campbell Memorial, Locus and World Fantasy Awards.

FRED SCHARMEN teaches architecture and urban design at Morgan State University. He is the co-founder of the Working Group on Adaptive Systems, an art and design consultancy based in Baltimore. His first book, *Space Settlements* – on NASA's 1970s proposal to construct large cities in space for millions of people – was published in 2019.

ANNA TALVI graduated from the Royal College of Art, School of Design in 2018. Her development of next-generation spacesuits is informed by symbiotic research across sartorial design, smart materials and biomedical engineering.

INDEX

A

A Signal from Mars (Taylor) - 18
Abramov, Oleg - 135
Adams, Constance - 121
adhocitecture - 184
Aelita: Queen of Mars (film) - 18, 43, 49
AI Space Factory, MARSHA - 130
alcohol consumption - 72
Aldrin, Buzz - 93
Alexander, Christopher, *A Pattern Language* -
    182
All-Story magazine - 42
Amazing Stories magazine - 41, 130
American Astronomical Society, *The Case for
    Mars* - 182
Ames Research Centre (NASA) - 184, 187
Antagonist Exo Muscle bodysuit - 94, 95
Antarctica - 181, 183, 184, 185
Apis Cor - 130
Apollo missions
    Apollo 13 - 69
    food - 118
    spacesuits - 93, 94
Ares, Greek god of war - 23
Arizona, University of
    Biosphere 2 programme - 166-9
    Dunes in Aonia Terra - 138
    Mars - 182
Ashley, Mike - 17-20
Asimov, Isaac, *The Martian Way and
    other stories* - 47
AstroPlant - 176
Atlas of Mars (NASA) - 182
Authentic Science Fiction Monthly - 20, 46
Avedissian, Beau - 161, 162

B

Balashova, Galina - 110-13
Bernal, John Desmond - 190
Bernal Sphere - 187, 190-3
BIOS-3 - 163-5
Biosphere 1 - 158
Biosphere 2 - 158, 166-9
Birch, Paul - 182
Bjarke Ingels Group, Mars Science City -
    196-7
Blue Book magazine - 42
Blue Mars (Robinson) - 19, 181, 182
Bogdanov, Alexander - 18, 181, 183
Bonestell, Chesley - 19, 78-81, 130, 181
boots - 107
Bova, Ben - 20
Bowie, David, 'Life on Mars?' - 19
Brackett, Leigh - 18
Bradbury, Ray - 183
    The Martian Chronicles - 19
Bradley, Marion Zimmer - 18
Brazil, Liana - 200
Brecht, Bertolt - 185

Bumper rockets - 75
Bursac, Aleksandr - 139

C

'The Call of Another World' - 19
calyx krater pot - 21
Campi, Ilaria - 139
'canals' on Mars - 37, 205
Carr, Michael R., *The Surface of Mars* - 182
Carter, Lin - 18
The Case for Mars (Zubrin) - 7
Cassini, Giovanni Domenico - 17
    *Observations in Bologna of the Rotation
    of Mars Around its Axis* - 27
Cellarius, Andreas, *Scenographia systematis
    Copernicani (Scenography of the
    Copernican world system)* - 24-5
CETEX (Committee on Contamination
    by Extraterrestrial Exploration) - 157
chairs - 71
Cheang, Ka Hou - 161, 162
'The Chessmen of Mars' (Rice Burroughs) - 42
China National Space Administration (CNSA),
    Chang'e 4 lunar lander - 173
CIMON (Computer Interactive Mobile
    Companion) - 124
Ciokajlo, Liz - 107
City of Cats (Lao She) - 18-19
Clark Telescope - 36, 39
Clarke, Jonathan - 133
closed life-support systems - 157-67
    BIOS-3 - 164-5
    closed-loop systems - 161-7
clothing - 93-4
    and 'Case for Mars' conferences - 182
    excursion suits - 131
    indoor - 11-12
    Mars boots - 107
    NEW HORIZONS collection (RÆBURN) - 154-5
    spaceplugs - 95
    spacesuits - 93-4, 97-106, 109
    on the Starship - 72
Clouds AO (Clouds Architecture Office),
    Mars Ice House - 142-3
coffee cups - 119
Cohen, Marc - 120
Collier's magazine - 130
    'Man will Conquer Space Soon!' series -
    78-81
Collier's Weekly - 19
colonisation of Mars - 10, 181, 183-5, 205
    closed-life support systems - 157-67
    habitats - 129-55
    terraforming - 19-20, 182, 183, 184-5,
    199, 205-7
colour image of Mars - 62
Conklin, Patricia 'Patsy' - 55
contamination - 157-60, 207
Copernicus, Nicolaus - 24, 26
Corrêa, Alvim - 40
Crete - 182

Cullen, Patricia - 19
cuneiform tablets - 21
Curiosity Rover - 17, 58-61, 63
'Cyborgs and Space' - 93
Cylindropuntia imbricata - 204

D
Davis, Donald E - 187, 191, 194
Davis, Stewart - 139
Deimos - 57
Destination Earth (cartoon) - 45
Dick, Philip K. - 183
Djkanovic, Ivan - 139
Dragon spacecraft - 89
Dresden Codex - 22
Dubai, Mars Science City - 196-7
Dunhuang star chart - 22
dying world view of Mars - 18-19, 205

E
ecopoiesis - 184
ecosystems, closed - 157-67
EDEN-ISS, Future Exploration Greenhouse
    (FEG) - 172
electricity supplies - 8-9
emergency drills, SpaceX Starships - 73
energy supply - 8-9
Ermacora, Thomas - 200
European Space Agency (ESA) - 8, 170, 207
Excelsior gondola - 98
exercising in space - 71, 95
ExoMars Rovers - 63

F
Fahandezh, Dorsa - 161, 162
Falcon - 88, 90
    Big Falcon Rocket - 90
'A Fighting Man of Mars' - 42
First World War - 18
fission power system - 8
Flammarion, Camille
    Astronomie Populaire (*Popular Astronomy*) -
    34
    Les Merveilles Céleste (*The Wonders of
        the Heavens*) - 34
    La Planète Mars (*The Planet Mars*) - 36
    telescope - 35
Fontenelle, Bernard de - 17
food and drink
    'Life on Mars: From Feces to Food' -
    161-3
    spacefood - 117, 118-19
Foster + Partners - 130
Mars Habitat - 136-7
Freitag, Captain Robert - 160
Friedrich, Caspar David, *The Wanderer above
    the Sea of Fog* - 159

G
Galileo - 17, 20
German Aerospace Center (DLR) - 124

German Spaceflight Society - 74
Gernsback, Hugo - 19
Ginsberg, Alexandra Daisy, *The Wilding of
    Mars* - 10, 184, 202-7
Goble, Warwick - 17, 18
Goodrich XH-5 spacesuit - 97
Grcic, Konstantin, dining table - 11, 122-3
Green Mars (Robinson) - 19, 181
greenhouses
    BIOS-3 - 165
    EDEN-ISS Future Exploration Greenhouse -
    172
    HASSELL 3D-printed habitat - 152
GrowStack vertical farm - 174
Guidice, Rick - 186, 187, 190, 192, 194

H
Harris, Gary L - 102
HASSELL - 140
    3D-printed habitat - 10, 148-53
Hawaii University, HI-SEAS habitat - 135
Hawking, Stephen - 10
Haynes, Robert - 184
Herschel, William - 17
    reflecting telescope - 28
heterotopia view of Mars - 207
HI-SEAS habitat - 135
Himmelskibet (*A Trip to Mars*) (film) -
    18, 19
HiRISE camera, images of the surface of Mars -
    138
Hoad, Lara - 139
Hooper, Tobe - 50
Huang, Jialu - 161, 162
Huygens, Christiaan - 17, 26
hydroponic farming systems - 174-5

I
ILC Dover spacesuits - 93, 94
Indian Space Research Organisation (ISRO) -
    62
inflatable habitats - 10, 148-53
InSight lander - 64
Institute of Isolation (film) - 109
International Space Station (ISS) -
    70, 71, 72
    clothing - 94
    food and drink - 119
    habitation - 119, 120-1
    vegetable production system 'VEGGIE' -
    171
Invaders from Mars (film) - 18, 44, 50
ISRO (Indian Space Research Organisation) -
    62
ISRU (in situ resource utilisation) - 9
ISS *see* International Space Station (ISS)

J
Je sais tout (*I know all*) magazine - 18, 19
Jones, William, orrery - 29

K

Kaku, Michio - 10
Kallipoliti, Lydia - 157-60, 161, 162, 163
Kepler, Johannes, *Astronomia Nova*
    (New Astronomy) - 26
Kimball, Ward - 76
Kittinger, Joseph - 98
Klushantsev, Paul - 49
Korolev, Sergei - 111

L

Lang, Fritz, *Metropolis* - 130
Lanos, Henry - 19
Lao She, *City of Cats* - 18-19
Lasswitz, Kurd - 183
lava - 57
Le Corbusier - 130
    Modular Man - 131
    Towards a New Architecture - 129
León, Pablo de - 102
Ley, Willy, and von Braun, Wernher, *Project*
    Mars - 78, 81
Liais, Emmanuel - 17
Life magazine - 130
'Life on Mars: From Feces to Food' - 161-3
life on Mars
    'canals' - 37, 205
    and colonisation - 9-10, 183, 207
    NASA on - 182-3
    in science fiction - 17-19
'Life on Mars?' (Bowie) - 19
Living Pod - 158-9
Lockheed Martin, Mars Base Camp - 86, 87
Loewy, Raymond - 70, 114-16
Lousma, Jack R - 117
Lowell, Percival - 17, 18, 36, 183, 205
    colour drawing of Mars - 39
    Mars - 17, 37

M

McCaughrean, Mark - 207
McKay, Chris - 184
McRae, Lucy
    'Astronaut Aerobics' - 108
    Institute of Isolation - 109
Man in Space (film) - 76
Mangala, Hindu god of Mars (painting) - 23
Mangalyaan (Mars Orbiter Mission probe) - 62
maps
    Dunhuang star chart - 22
    of Mars - 30-1, 33, 34, 37
Mariner 4, photographs of the Martian
    surface - 17, 52-4
Mars 2100 - 200-1
Mars Analog Research Station Program - 133
Mars Attacks! (film) - 18
Mars Base Camp - 86
    lander - 87
Mars (Bova) - 20
Mars City Design and Alpha Team - 139
Mars Excursion Module (MEM) - 157

Mars (film) - 49
Mars Ice House - 140-3
Mars (Lowell) - 17, 37
The Mars Project (von Braun) - 19
Mars Science City - 196-7
Mars Society - 7, 133-4
Mars Underground - 182, 184
Mars-3 lander - 54
MARSHA - 130
The Martian Chronicles (Bradbury) - 19
The Martian (film) - 159
The Martian Way and other stories (Asimov) -
    47
The Martian (Weir) - 20
Marvel Comics - 130
Mead, Syd - 187
MELiSSA (Micro-Ecological Life Support
    System Alternative) - 170
MEM (Mars Excursion Module) - 157
MER-2 Rover - 56
Mercury space capsule - 99
    spacesuits - 99
Mesiha, Mariam - 161, 162
Messeri, Lisa, *Placing Outer Space: An*
    Earthly Ethnography of Other Worlds -
    132
Metropolis (film) - 130
Micro-Ecological Life Support System
    Alternative (MELiSSA) - 170
minerals - 9
Mir space station - 112, 113
Mission to Mars (Moore) - 19
Modernist architecture - 129, 130, 131-2
Montalti, Maurizio - 107
Moon landings - 11, 69, 75
    spacesuits - 93, 94
    moons of Mars - 19, 57
Moore, Patrick - 19
movement of Mars - 38
Mulyani, Vera - 139
Musk, Elon - 10, 71, 72, 91
    plans for building on Mars - 131, 182,
        183, 207
    Tesla - 9
    see also SpaceX

N

NASA
    Ames Research Centre - 184, 187
    Apollo 13 - 69
    Atlas of Mars - 182
    Bumper rockets - 75
    Curiosity Rover - 17, 58-61, 63
    fly-by photograph of the surface of Mars -
        17
    Habitability Study - 114-15
    InSight lander - 64
    Living Pod - 158-9
    Manned Space Flight Center - 160
    Mars mission - 8, 69, 121
    Office of Technology Utilization - 157

Orion Multi-Purpose Crew Vehicle -
    8, 84, 85
Saturn V rocket - 69, 75, 77
Space Launch System (SLS) - 69, 84, 85
Systems Engineering Handbook - 93
3D-Printed Habitat Challenge -
    130-1, 132, 140
Transhab prototype - 121
Viking probes - 182
NDX-1 spacesuits - 102-3
Nebo Zovyot (The Sky is Calling) - 49
Newman, Dava - 104-5
Nixon, David - 120
nuclear-fisson reactors - 9

O
O'Callaghan, Eckersley - 130
O'Neill Cylinders - 187, 194-5
O'Neill, Gerard - 187, 190
Opportunity Rover - 17, 183
Orbiter Mission probe (Mangalyaan) - 62
Orion Multi-Purpose Crew Vehicle - 8, 84, 85
orrery (mechanical model of the solar
    system) - 29
oxygen
    and closed life-support systems -
        157, 158
    producing and storing - 8-9

P
Park, Jasmine - 139
Paterson, Katie, *Timepieces (Solar System)* -
    156
Pathfinder - 56
Paul, Frank R - 41, 130
Pearson's Magazine - 17, 18, 41
Phobos - 19, 57
planetary contamination - 157-60, 207
planetary imagination - 132
plants
    AstroPlant - 176
    Ginsberg's *The Wilding of Mars* - 10, 184,
        202-7
    hydrophonic farming systems - 174-5
    vegetable production systems - 171-2
    the wilding of Mars - 10, 184, 202-7
A Princess of Mars - 18
Proctor, Richard A., *Other worlds than ours* -
    30
Project Excelsior - 98
Protazanov, Yakov - 43
psychological wellbeing - 11, 109

R
RÆBURN, 153-5
realist views of Mars - 19
Reconnaissance Rover - 138
recycling
    clothing - 155
    waste - 158, 159, 160
red colour of Mars - 17, 18, 20, 23

Red Mars (Robinson) - 19, 181, 182
Red Star (Bogdanov) - 18
reflecting telescope - 28
regolith (Martial topsoil) - 5, 11
    as habitat shield - 139, 148
Return to Mars (Bova) - 20
Rice Burroughs, Edgar - 18, 42, 183
Rive, Russ - 200
Robinson, Kim Stanley - 5, 19, 181-5
robots
    as builders on Mars - 136
    CIMON - 124
The Rockets of Human Spaceflight (Skrabek) -
    82-3
'Rocknest' site - 58
rocks, Martian - 19-20
    'Mazatzal' - 57
romantic view of Mars - 18-19

S
Sagan, Carl - 9-10, 71, 207
    The Cosmic Connection - 19
Sattelmeier, Paul - 182
Saturn as Seen from Titan (Bonestell) - 78
Saturn V rocket - 69, 75, 77
Scharmen, Fred - 5
Schiaparelli, Giovanni - 17, 32, 36, 205
    *Osservazioni di Marte (Observations of
        Mars)* - 30-1, 33
science fiction - 17-19, 49, 130, 181, 183
'The Scientific Adventures of Baron
    Münchausen' - 19
scientific research on Mars - 181, 183, 200
Scott, Ridley - 159
SEArch + (Space Exploration Architecture)
    - 130
    Mars Ice House - 140-3
    Mars X-House - 12, 144-7
Second World War - 19
Shelton, Richard - 47
Shepherd, Alan - 99
Skrabek, Tyler, *The Rockets of Human
    Spaceflight* - 82-3
Skylab - 115-17, 118
sleeping in space - 71
Slipher, EC - 38
SLS (Space Launch System) - 69, 84, 85
Sojourner Rover - 56, 183
solar panels - 9
solar radiation - 11
Soleri, Paolo - 182
southern polar ice caps - 17
Soviet science fiction - 49
Soviet space missions - 110-13
    BIOS-3 programme - 163-5
    Mars programme - 54
Soyuz LOK - 110-11, 112
Space Launch System (SLS) - 69, 84, 85
Space Shuttle - 84
SpaceFactory, MARSHA - 130
spacesuits - 93-4, 97-106, 109

BioSuit™ - 104-5
Mercury - 99
NDX-1 spacesuits - 102-3
SpaceX - 106
SpaceX - 91, 207
   Big Falcon Rocket (BFR), 90
   booster rockets - 7, 9
   development of a Mars city - 198
   Dragon spacecraft - 89
   Falcon booster, rockets - 88, 90
   Mars settlement with Starship - 198
   Merlin vacuum engine - 88
   spacesuit - 106
   Starship - 70-2, 91
   terraforming Mars - 199
Spirit Rover - 56-7, 183
Stanford Torus - 186-7, 188-9
Starship - 70-2, 91
   landing on Mars - 92
Stepanova, Anastasiya - 133
Stockbridge Technology Centre, GrowStack
   vertical farm - 174
the Sun, Mars orbits of - 26, 64
SuperUber, *Mars 2100* - 200
surface of Mars
   digital image of - 52-3
   HiRISE camera images - 139
   photographs of - 17, 54, 55
Sutherland, John - 45

T
tables - 70, 120
   Grcic's dining table - 11, 122-3
Talvi, Anna - 11-12, 93-5
Taylor, Raymond, *A Signal from Mars* - 18
*Tekhnika Molodezhi (Technology for the
   Youth)* - 48
telescopes - 17
   Bardou refractor telescope - 36
   Clark Telescope - 36, 39
   drawing of an aerial telescope - 26
   reflecting - 28
   surface of Mars - 31
temperatures on Mars - 8
terraforming Mars - 19-20, 182, 183, 184-5,
   199, 205-7
3D-Printed Habitat Challenge - 130-1, 132,
   140
'Thuvia, Maid of Mars' (Rice Burroughs) - 42
Timepieces (Solar System) - 156
Tolstoi, Alexander, *Aelita* - 49
The Torus Wheel - 186
Total Recall (film) - 50-1
Transhab prototype - 121
travel time to Mars - 12
A Trip to Mars (film) - 18
Trotti, Guillermo - 104
Tsarev, Y - 49
Tsiolkovsky, Konstantin, *Album of Cosmic
   Journeys* - 73

U
'Under the Moons of Mars' - 18
United Nations, International Outer Space
   Treaty - 207
Urbano, Carl, *Destination Earth* - 45

V
Valles Marineris hemisphere of Mars - 7
Verhoeven, Paul - 50, 51
Vezzuli, Paolo - 139
Viking 1 Lander - 55
von Braun, Wernher - 74, 75, 76, 77
   Das Marsprojeckt (The Mars Project) - 19,
     69, 70
   and Ley, Willy, *Project Mars* - 78, 81
   voyage to Mars - 5, 11, 69-124

W
The Wanderer above the Sea of Fog
   (Friedrich) - 159
The War of the Worlds (Wells) - 17-18, 40-1
warlike view of Mars - 18
'Warlord of Mars' (Rice Burroughs) - 42
waste, and closed life-support systems -
   157-8, 159, 160
Watney, Mark - 159
Weir, Andy, *The Martian* - 20
Welles, Orson - 18, 41
Wells, HG - 183
   The War of the Worlds - 17-18, 40
Wells, Isabella - 161, 162
The Wilding of Mars (Ginsberg) - 10, 184,
   202-7
windows, SpaceX Starships - 70-1
women and missions to Mars - 94
Wonder Stories magazine - 46

X
X-House - 144-7
X-House V2 - 130
X.Earth olfactory gloves-mask - 94, 95

Y
Yutu-2 Moon Rover - 173

Z
zero gravity - 5
Zubrin, Robert - 133
   The Case for Mars - 7

PICTURE CREDITS

Every reasonable attempt has been made to identify owners of copyright. Errors and omissions notified to the publisher will be corrected in subsequent editions.

Abbreviations are: r-right, l-left, t-top, b-bottom, c-centre.

Cover image ESA/DLR/FU Berlin, licensed under CC BY-SA 3.0 IGO, https://sci.esa.int/web/mars-express/-/37808-nicholson-crater-on-mars (original image altered)

Adler Planetarium, Chicago, Illinois: p.28, p.30 (t)
Alamy Stock Photo/Chronicle: p.18 (t)
Alexandra Daisy Ginsberg. Additional research and development: Ness Lafoy, Johanna Just, Ioana Man and Stacie Woolsey. Software development/simulation design: Tom Betts/Nullpointer, Jelena Viskovic and Ana Maria Nicolaescu. Sound: Sam Conran. With thanks: Professor Luis Campos, Baruch S. Blumberg NASA/Library of Congress Chair in Astrobiology 2016-17; Dr Lynn Rothschild, NASA; Professor James Head, Brown University; Lisa White. Commissioned by the Vitra Design Museum and the Design Museum, London. With support from Cité du Design, Saint-Étienne: pp.202-3, p.204, pp.206-7
AI SpaceFactory/Plomp: p.130, p.131
Allen Family Collection: p.74 (cl, cr, b)
Photos Archive of the Kirensky Institute of Physics, Krasnoyarsk Research Center, Siberian Branch of Russian Academy of Sciences: p.164, p.165
© Archive of the Russian Academy of Sciences: p.73
AstroPlant/Gernot Kuenzel: p.176
A/V Geeks/National Film Preservation Foundation, San Francisco: p.45
Biblioteca comunale di Trento: pp.24-5
BIG - Bjarke Ingels Group. Client: Government of United Arab Emirates. Collaborators: Mohammed bin Rashid Space Centre, Dubai Municipality, BIG Landscape, BIG Ideas, BIG Engineering: p.196, p.197
© Bonestell LCC: p.78 (b), p.79 (tl, tr, br), p.81 (t, b), p.181
Courtesy of Bonhams: p.35
British Library: p.22 (b)
© Trustees of the British Museum: p.21, p.23
CELLS Research Group at Limerick IT: p.172
Center of Space Exploration, Ministry of Education (COSE): p.173 (b)
China National Space Administration: p.173 (t)
Constance Adams Archives: p.121 (tl, tr, bl)

Danish Film Institute: p.19 (b)
Dibner Library of the History of Science and Technology, Smithsonian: p.26 (t)
DLR/T. Bourry/ESA: p.124
Photo Dougas Sonders: p.104
Dovzhenko Film Studios/Ukrainian Rights Management Group: p.49 (b)
ESA/DLR/FU Berlin, licensed under CC BY-SA 3.0 IGO, https://sci.esa.int/web/mars-express/-/55576-kasei-valles-mosaic (original image altered): p.13, p.16
ESA/DLR/FU Berlin (G. Neukum), licensed under CC BY-SA 3.0 IGO, http://www.esa.int/spaceinimages/Images/2010/10/Melas_Chasma_on_Mars (original image altered): p.125, p.128
Photo Francesc Melcion/Diari Ara: p.170
Images used with the acknowledgment of the Frank R. Paul Estate: p.41, p.46 (bl, br)
FUTURE SYSTEMS (David Nixon and Jan Kaplicky): p.120 (t, bl)
FUTURE SYSTEMS (David Nixon and Jan Kaplicky)/Photo NASA: p.120 (br)
Galina Balashova Family Archive/Collection Philipp Meuser: pp.110-1, p.112, p.113
Photo George Ellsworth: p.107
Getty Images/Benjamin Rasmussen: p.134 (t)
Getty Images/Bert Lawson/Toronto Star: p.18 (b)
Getty Images/Bettmann: p.98 (l)
Getty Images/Harold M. Lambert: p.99 (b)
Getty Images/Leon Neal/AFP: p.63
Getty Images/Pallava Bagla/Corbis: p.62 (t)
Getty Images/Science and Society Picture Library: p.37 (b)
Courtesy of Gosfilmofond: p.49 (tr, tl, cl, cr)
Growstack/Growing Underground www.growing-underground.com: p.174
HASSELL and Eckersley O'Callaghan: p.10, pp.148-9, pp.150-1, p.152, p.153
Heritage Auctions, HA.com: p.40, p.42 (tl, tc, tr), p.44, p.47, p.81 (t)
INAF-Astronomical Observatory of Brera, Milan: p.30 (b), p.31, p.33
Photo courtesy of the Institute of Ecotechnics: p.166 (t)
IRPI LLC: p.119 (tl)
ISRO/ISSDC/Justin Cowart: p.62 (b)
Photo Jenni Pereyda: pp.168-9
© Katie Paterson 2019. Image courtesy the artist and James Cohan, New York. Photo © John McKenzie: p.156
Lavazza: p.119 (tr)
Leiden University Libraries, ms. HUG 5, fol.41r: p.26 (bl)
Leiden University Libraries, ms. HUG 28, fol.212r: p.26 (br)
Library of Congress Prints and Photographs Division Washington, D.C.: p.74 (t);
Library of Congress, Prints and Photographs

Division, Washington, D.C. Carol
M. Highsmith's America: p.166 (b)
Lockheed Martin: p.86, p.87
Lowell Observatory Archives: p.36, p.37 (t),
p.38, p.39
Lucy McRae: p.108, p.109 (bl, br)
Lucy McRae/Daniel Gower: p.109 (tr)
Lucy McRae/Lotje Sodderland: p.109
(tl, cl, cr)
The Mars Society: p.133 (t, bl), p.134 (b)
© Metro-Goldwyn-Mayer: p.50 (t)
Image courtesy of Museo Astronomico di
Brera. © Archivio storico Fondazione
Corriere della Sera: p.32
The Museum of Cosmonautics, Moscow: p.54 (b)
NASA: p.8, p.9, p.69, p.71 (b), p.75, p.77,
p.84, p.85, p.93, p.99 (t), p.116 (b),
p.117, p.118, p.119 (b), p.121 (br),
p.157, p.205
NASA Ames Research Center: p.186, p.187,
pp.188-9, p.190, p.191, pp.192-3, p.194,
p.195
NASA/Drawing Matter: p.114 (tl, bl, br)
NASA/Heritage Auctions, HA.com: p.115,
p.116 (t)
NASA/JPL: p.66 (t), p.66 (tl), p.57 (t)
NASA/JPL-Caltech: p.7, pp.52-3, p.55 (b),
p.58 (b), p.64
NASA/JPL-Caltech/Cornell/USGS: p.57 (c)
NASA/JPL-Caltech/Dan Goods: p.17 (t)
NASA/JPL-Caltech/MSSS: pp.58-9 (t),
p.59 (b), pp.60-1
NASA/JPL-Caltech/Univ. of Arizona: p.138
NASA/JPL-Caltech/Univ. of Arizona, PIA12289
(original image altered): pp.14-5
NASA/JPL-Caltech/Univ. of Arizona, PIA12178
(original image altered): p.65, p.68
NASA/JPL-Caltech/Univ. of Arizona, PIA21551
(original image altered): p.66-7
NASA/JPL-Caltech/Univ. of Arizona, PIA22331
(original image altered): pp.126-7
NASA/JPL-Caltech/Univ. of Arizona, PIA11178
(original image altered): p.177, p.180
NASA/JPL-Caltech/Univ. of Arizona, PIA12260
(original image altered): p.178-9
NASA/JPL/Cornell: p.56 (b)
NASA/JPL/Cornell/Texas A&M: p.57 (b)
NASA/JPL/KSC: p.56 (tr)
NASA/Kim Shiflett: p.171
NASA/University of North Dakota, p.102,
p.103
National Air and Space Museum Archives:
p.97, pp.100-1
© National Maritime Museum, Greenwich,
London: p.29
National Museum of the US Air Force/Volkmar
Wentzel: p.98 (r)
Ottilie Landmark: p.94, p.95
Saxon State and University Library Dresden:
p.22 (t)
Science Photo Library/Volker Steger: p.105

SEArch+ (Space Exploration Architecture):
p.12, pp.144-5, p.146, p.147
SEArch+ (Space Exploration Architecture)/
Clouds AO (Clouds Architecture Office):
pp.140-1, p.142, p.143
SPACE10/Nicklas Ingemann: p.175 (t)
SPACE10/Rory Gardiner: p.175 (b)
SpaceX: p.70, p.71 (t), p.88, p.89, p.90,
p.91, p.92, p.106, p.198, p.199
Special Media Archives Services Division,
National Archives, College Park, MD:
p.158
STUDIOCANAL: p.50 (b), p.51
Tekhnika Molodezhi: p.48
Creative Directors and Executive Producers:
Thomas Ermacora, Liana Brazil and Russ
Rive (SuperUber); Script Writer:
Alexandre Rossi; Lead Animator: Bernardo
Leite; Illustrator: Andres Ramos;
Animators: Rodrigo Borges, Rodrigo
Mantega; Sound Design and Music: Nick
Ryan Studio; Voice Narrators: Trine
Garrett, John Kapansa; Director of
Operations: Fabiano Martins; Director of
Technology: Carlos Oliveira; Producer:
Mariana Ferman; Project Manager: Liana
Markus; Technical Sound Design: Davey
Williamson; Technical coordination: Blue
Elephant: pp.200-1
The Martian © Twentieth Century Fox.
All rights reserved: p.159
Tyler Skrabek: pp.82-3
Courtesy of The University of Arizona: p.167
University of Hawai'i at Mānoa. Photo Angelo
Vermeulen: p.135 (t)
University of Hawai'i at Mānoa. Photo Oleg
Abramov: p.135 (b)
University of Oklahoma Libraries, History of
Science Collections: p.27
Yusuke Murakami/The Mars Society: p.133 (br)

ACKNOWLEDGEMENTS

This book was published in conjunction with the exhibition *Moving to Mars* at the Design Museum, London, 18 October 2019 to 23 February 2020.

Curators: Andrew Nahum and Eleanor Watson

Chief Curator: Justin McGuirk

Curatorial Research Assistants: Domenica Bates, Margherita Manca and Anna Xene Marchant

Exhibition Project Managers: Zoe Few and Cleo Stringer

Exhibition Coordinator: Jessica Taylor

Exhibition Design: All Things Studio

Exhibition Graphic Design: Fabrique

Visitor Experience Design: NorthernLight

The Design Museum owes its gratitude to all the lenders of the exhibition. Special thanks to colleagues and friends who have generously shared their knowledge and advice to aid the curatorial development of the exhibition.

The Design Museum would also like to thank the following organisations and people: AI SpaceFactory, Allen Family Collection, Grant Anderson, Mike Ashley, Galina Balashova, David Barraclough, Debra Bivens, Liana Brazil, Brera Observatory, The British Library, The British Museum, Rebecca Carpenter, China National Space Administration, Marc Cohen, Pablo de Leon, Drawing Matter, Maria Joao Durao, Simon Dwyer, Thomas Ermacora, European Space Agency, Farnborough Air Sciences Trust, Foster + Partners, Manuel Jimenez Garcia, Alexandra Daisy Ginsberg, Konstantin Grcic, GrowStack, Sanjeev Gupta, Rick Guidice, HASSELL, Sandra Hauplik-Meusburger, Horden Cherry Lee Architects, ILC Dover, Anab Jain, Richard Jurek, Lydia Kallipoliti, University of Leiden Library, Lightfield London, Lucy McRae, Gregory Mueller, NASA Ames Research Center, NASA Jet Propulsion Laboratory, David Nixon, Officina Corpuscoli, Sean O'Mara, Katie Paterson, RÆBURN, Russ Rive, The Royal Asiatic Society, Royal Museums Greenwich, David Meerman Scott, SEArch+, John Sisson, SpaceX, Franziska Steingen, Anna Talvi, Andreas Vogler, Kim Wallace and Wieger Wamelink.

Design Museum Publishing
Design Museum Enterprises Ltd
224-238 Kensington High Street
London W8 6AG
United Kingdom

designmuseum.org

First published in 2019
© 2019 Design Museum Publishing

ISBN 978 1 872005 46 1

Publishing Manager: Mark Cortes Favis
Assistant Editor: Mary Miller
Picture Researcher: Nikos Kotsopoulos
Photo Editor: Tim Blann
Copyeditor: Simon Coppock
Proofreader. Ian McDonald
Designers: Studio Julia, julia.studio

Many colleagues at the Design Museum have
supported this book, and thanks go to them all.

Distribution

UK, Europe and select territories around
the world
Thames & Hudson
181A High Holborn
London WC1V 7QX
United Kingdom
thamesandhudson.com

USA and Canada
ARTBOOK | D.A.P.
75 Broad Street, Suite 630
New York, NY 10004
United States of America
artbook.com

Printed and bound in Malta by Gutenberg Press